IET TELECOMMUNICATIONS SERIES 87

Real Time Convex Optimisation for 5G Networks and Beyond

Other volumes in this series:

Real Time Convex Optimisation for 5G Networks and Beyond

Long D. Nguyen, Trung Q. Duong, and
Hoang Duong Tuan

The Institution of Engineering and Technology

Published by The Institution of Engineering and Technology, London, United Kingdom

The Institution of Engineering and Technology is registered as a Charity in England & Wales (no. 211014) and Scotland (no. SC038698).

The Institution of Engineering and Technology
Michael Faraday House
Six Hills Way, Stevenage
Herts, SG1 2AY, United Kingdom

www.theiet.org

British Library Cataloguing in Publication Data
A catalogue record for this product is available from the British Library

ISBN 978-1-78561-959-5 (hardback)
ISBN 978-1-78561-960-1 (PDF)

Typeset in India by Exeter Premedia Services Private Limited
Printed in the UK by CPI Group (UK) Ltd, Croydon

Contents

About the Authors

Long D. Nguyen is a Lecturer at Dong Nai University and Adjunct Assistant Professor at Duy Tan University, Vietnam. His research interests include convex optimisation techniques for resource management in wireless communications, energy efficiency approaches for 5G networks (heterogeneous networks, relay networks, cell-free networks, and massive MIMO) and real-time optimisation for wireless communication networks and Internet of Things. He holds a PhD in Electronics and Electrical Engineering from Queen's University Belfast, UK.

Trung Q. Duong is a professor at Queen's University Belfast, UK, and a Research Chair of Royal Academy of Engineering. His research interests include wireless communications, signal processing, machine learning and optimisation for wireless networks. He serves as editor for IEEE Trans on Wireless Communications and executive editor for IEEE Communications Letters. He received the Royal Academy of Engineering Research Fellowship (2016-2020) and won the Newton Prize in 2017. He is co-editor of the IET book Trusted Communications with Physical Layer Security.

Hoang D. Tuan is a professor at the School of Electrical and Data Engineering, University of Technology Sydney, Australia. He has been involved in research on optimisation, control, signal processing, wireless communication, and biomedical engineering for more than 20 years. He received his PhD in Applied Mathematics from Odessa State University, Ukraine.

Chapter 1

Convexity and convex optimisation problems

1.1 Convex sets

A set χ that contains the segment connecting any two points of it is called a convex set [1], i.e. is χ a covex set if and only if for any $x_1, x_2 \in \chi$ and $\theta \in [0, 1]$, is it true that

$$\theta x_1 + (1-\theta) x_2 \in \chi. \tag{1.1}$$

Figure 1.1 provides some examples of convex set and non-convex set.

Some simple convex sets are provided in Table 1.1 and Figure 1.2.

In Table 1.2, we introduce examples of convex sets based on the intersections of convex sets. In particular, the intersection of linear subspaces, affine subspaces or convex cones is also a convex set.

1.2 Convex functions

A function $f(x)$, as shown in Figure 1.3, is said to be convex in a convex domain of $\mathrm{dom} f$ if the following Jensen's inequality holds for all $x, y \in \mathrm{dom} f$ and $\theta \in [0, 1]$

$$\theta f(x) + (1 - \theta) f(y) \geq f(\theta x + (1 - \theta) y). \tag{1.2}$$

If $-f(x)$ is convex then $f(x)$ is said to be concave, i.e. the following inequality holds for all $x, y \in \mathrm{dom} f$ and $\theta \in [0, 1]$

$$\theta f(x) + (1 - \theta) f(y) \leq f(\theta x + (1 - \theta) y). \tag{1.3}$$

An affine function $f(x)$ is both convex and concave and thus satisfies the following condition

$$\theta f(x) + (1 - \theta) f(y) = f(\theta x + (1 - \theta) y). \tag{1.4}$$

Based on graphs, Figure 1.4 illustrates a convex function (a), a concave function (b) and a function which is neither convex nor concave (c).

Some examples of convex and concave functions are provided in Tables 1.3 and 1.4 such as affine functions, convex quadratic functions, norm functions and quadratic over linear. They are very popular in engineering. In addition, convex or concave functions are very important to recognise a convex optimisation problem

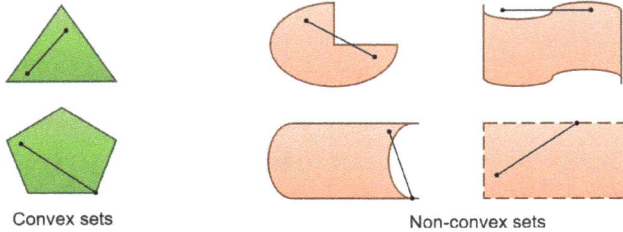

Convex sets   Non-convex sets

*Figure 1.1   (a) A green triangle and pentagon with their boundary are convex
sets, (b) A clock shape, the winding flag shape and a rectangle or
square with broken boundaries are not convex (non-convex) sets[1].*

(as discussed in Subsection 1.3). Depending on the characteristics of the mathematical optimisation models and applications, there are several classes of mathematical programming to convex optimisation such as linear programming (LP), quadratic programming (QP), second-order cone programming (SOCP) and semi-definite programming (SDP). For example, a simple linear program can consist of a linear objective, one or several linear constraints. The classes of convex programming will be introduced in the next chapter.

1.3   Convex optimisation problems

Over the past years, convex optimisation theory has proven its important role in engineering. In particularly, convex optimisation-based models are applied in numerous

Table 1.1   Examples of convex sets [1]

Convex set	Description
Cone (non-negative homogeneous)	$\{\theta_1 x_1 + \theta_2 x_2 \in \chi \mid x_1, x_2 \in \chi, \theta_1, \theta_2 \geq 0\}$, χ is the convex set
Polyhedron	$\{x \mid a_j^T x \leq b_j, j = 1, ..., m, c_j^T = d_{i,j} = 1, .., p\}$
Norm ball	$\{x \mid \|x - x_C\| \leq r\}$, $x \in \mathbb{R}^n, r > 0$ is radius, x_c is center point
Norm cone	$\{(x, t) \mid \|x\| \leq t\} \subseteq \mathbb{R}^{n+1}$
Hyperplane	$\{x \mid a^T x = b\}$, $a \in \mathbb{R}^n, a \neq 0, b \in \mathbb{R}$
Half-spaces	$\{x \mid a^T x \leq b\}$, $a \in \mathbb{R}^n, a \neq 0, b \in \mathbb{R}$
Euclidean ball	$\{x \mid \|x - x_c\|_2 \leq r\}$, $r > 0$ is radius, x_c is the center point
Ellipsoid	$\{X \mid (x - x_c)^T P^{-1}(x - x_c) \leq 1\}$, $P \in \mathbb{R}^{n \times n}$ is symmetric and positive definite, x_c is the center

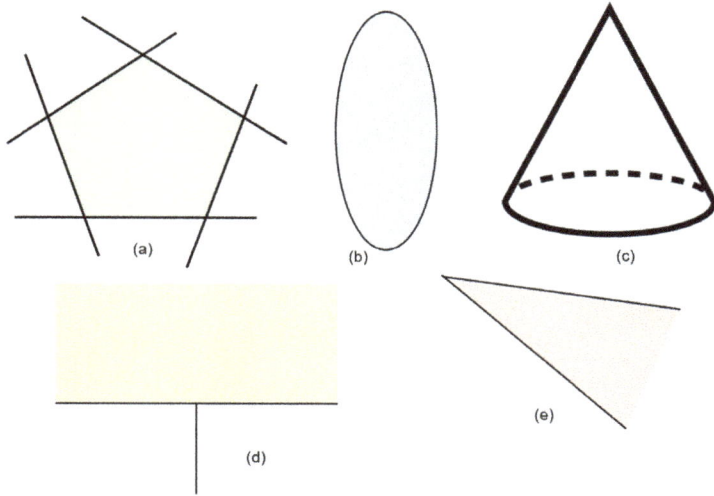

Figure 1.2   *Examples of convex sets: a polyhedron (a), a ellipsoid (b), a norm cone (c), a hyperplane ($a^T x = b$) $\in \mathbb{R}^2$ shape (d) and a pie slice (e).*

signal-processing and wireless communication scenarios. Together with the explosion of wireless communication, convex optimisation has become the most potential approach for the design, analysis and deployment of wireless communication systems. In fact, many aspects of wireless networks such as beamforming design, resource allocation, spectral and energy efficiency maximisation are exploited and addressed by convex optimisation. In addition, many non-convex optimisation problems in wireless networks can be solved by non-convex approaches or can be converted into convex ones which are handled using various convex optimisation algorithms.

There are many reasons for adopting convex optimisation in engineering. The major advantage is that a local optimal solution of a convex optimisation problem is also its global solution. In convex problems, a global optimal point is found by any locally optimal point based on the concept of convexity. Meanwhile, a local optimum is much more easy to find than a global optimum. When an optimisation problem (i.e. resource allocation problem in wireless communication system) is built as a convex

Table 1.2   *Some properties of convex sets*

Intersection of convex sets	Description
Linear subspace	$S = \{x \mid Ax = 0\}$
Affine subspace	$S = \{x \mid Ax = b\}$
Polyhedral set	$S = \{x \mid Ax \leq b\}$
Positive semidefinite cone	$S = \{x \mid x^T Ax \geq 0, A = A^T\}$
Second order cone	$S = \{(t, x) \mid t \geq \|x\|\}$

A "line segment" connecting f(x) and f(y):
$$\theta f(x) + (1 - \theta)f(y)$$

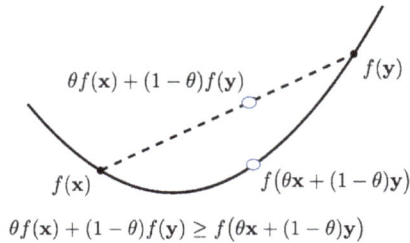

$\theta f(\mathbf{x}) + (1 - \theta)f(\mathbf{y})$

$f(\mathbf{y})$

$f(\mathbf{x})$

$f(\theta \mathbf{x} + (1 - \theta)\mathbf{y})$

$$\theta f(\mathbf{x}) + (1 - \theta)f(\mathbf{y}) \geq f(\theta \mathbf{x} + (1 - \theta)\mathbf{y})$$

Figure 1.3   Graph of a convex function.

problem (i.e. linear programming), the next task is to find its local optimum by an available optimisation algorithm. In addition, convex optimisation methods are capable of providing a good heuristic suboptimal solutions or semi-exact solutions to non-convex problems. For instance, interior-point algorithms are often highly efficient in solving optimisation problems with polynomial computational complexity. Further to this, a wide range of optimisation software available can only solve convex optimisation problems.

Meanwhile, there is a lack of efficient methods and software for solving non-convex problems at present. Even if they were available, there would be no guarantee that the optimal solution can be explicitly found. In contrast, convex optimisation methods often bring the insight of optimal solution models which are relevant to the original non-convex problems, i.e. signal processing and wireless communication application.

A feasible optimisation problem is to maximise or minimise a real function (objective function) by systematically choosing input values (variables) from an allowed set that satisfies all the constraint functions. A general convex optimisation problem is given as

$$\underset{x}{\text{minimise}} \quad f(x) \tag{1.5a}$$
$$\text{subject to} \quad g_i(x) \leq 0, \; i = 1, ..., I, \tag{1.5b}$$
$$h_j(x) = 0, \; j = 1, ..., J, \tag{1.5c}$$

where $x \in \mathbb{R}^n$ is the variable and $f: \mathbb{R}^n \to \mathbb{R}$ is the objective function. For $g_i : \mathbb{R}^n \to \mathbb{R}$ is the inequality constraints. For $h_j : \mathbb{R}^n \to \mathbb{R}$ is the equality constraints.

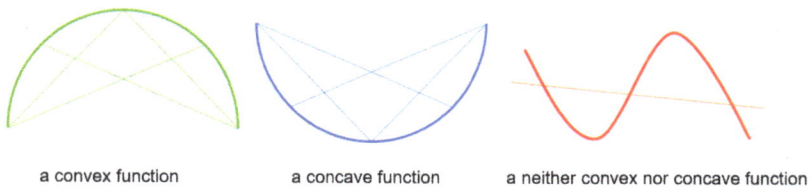

a convex function   a concave function   a neither convex nor concave function

Figure 1.4   A graph of a concave, convex and non-convex function.

Table 1.3   Examples of convex functions.

Convex functions	Formulation
Affine functions (on $\mathbb{R}^n$)	$f(x) = a^T x + b, x, a \in \mathbb{R}^n, b \in \mathbb{R}$
Affine functions (on $\mathbb{R}^{n \times m}$)	$f(X) = Tr(AX) + b, A, X \in \mathbb{R}^{n \times m}, b \in \mathbb{R}$
Convex quadratic functions (on $\mathbb{R}$)	$f(x) = ax^2 + bx + c, x, a, b, c \in \mathbb{R}, a > 0$
Quadratic functions (on $\mathbb{R}^n$)	$f(x) = x^T P x + 2q^T x + c, x, q \in \mathbb{R}^n, c \in \mathbb{R},$ P is positive semi-definite
Power functions	$\begin{cases} f(x)=x^p, \ p \geq 1 \text{ or } 0, \\ f(x)=e^{ax}, \ x, a \in \mathbb{R}_{++} \end{cases}$
Logarithm functions (on $\mathbb{R}_{++}$)	$\begin{cases} f(x)= \qquad -\log x, \\ f(x)=x \log x \text{ (negative entropy)} \end{cases}$
Log-sum-exp	$f(x) = \log \sum_{i=1}^{n} e^{x_i}, \forall x_i \in \mathbb{R}$
Norms	$\|x\|$, e.g. $\|\vec{x}\|_1, \|x\|_2, \|x\|_k, \ldots$
Quadratic over linear	$f(x,y) = \frac{x^2}{y}, \forall x, y \in \mathbb{R}_{++}$

The optimisation problem above is called convex if the objective function f and g_i are convex while h_i are affine.

   Example 1.6: Consider an example of convex problem in [1]:

$$\underset{x}{\text{minimise}} \quad f(x) = x_1^2 + x_2^2 \tag{1.6a}$$
$$\text{subject to} \quad g_1(x) = x_1 \leq 0, \tag{1.6b}$$
$$h_1(x) = x_1 + x_2 = 0. \tag{1.6c}$$

On the other hand, the below problem in (1.7) is not convex but both problems (1.6) and (1.7) have the same optimal solution($x_1^* = 0, x_2^* = 0$). Note that the inequality constraint (1.7b) is non-convex.

$$\underset{x}{\text{minimise}} \quad f(x) = x_1^2 + x_2^2 \tag{1.7a}$$

Table 1.4   Examples of concave functions

Concave functions	Formulation
Power	$f(x) = x^p, \ 0 \leq p \leq 1, x > 0$
Logarithm	$\log x, x > 0$
Logarithmic determinant	$\log \det X, X \succ 0, X \in \mathbb{R}^{n \times n}$
Geometric mean	$f(x) = (\prod_{i=1}^{n} x_i)^{\frac{1}{n}}, x_i > 0$

subject to $\tilde{g}_1(x) = x_1/(1 + x_2^2) \leq 0,$ (1.7b)

$\tilde{h}_1(x) = x_1 + x_2 = 0.$ (1.7c)

Example 1.8: An entropy maximisation problem with $x \in \mathbb{R}^N$ is given as follows:

$$\underset{x \geq 0}{\text{maximise}} \quad \sum_{i=1}^{N} x_i \log x_i$$ (1.8a)

$$\text{s.t.} \quad \sum_{i=1}^{N} x_i = 1,$$ (1.8b)

$$Ax = b,$$ (1.8c)

where the objective function (1.8a) is a concave function and (1.8b) and (1.8c) are linear constraints. Hence, the problem in (1.8) is a convex optimisation problem by maximising a concave function.

Example 1.10: Consider a wireless downlink transmission in which a base station is equipped with N antennas to serve K single-antenna users. The channel is modelled as

$$y_i = h_i^H x + z_i, \quad i = 1, ..., K$$ (1.9)

where $x \in \mathbb{C}^N$ denotes the transmit signal. The transmit signal is defined as $x = \sum_{i=1}^{K} v_i w_i$, where $\in \mathbb{C}^N$ is the information signal for the ith user and $w_i \in \mathbb{C}^N$ is the beamforming vector for the ith user. $h_i \in \mathbb{C}^{1 \times N}$ is the channel vector and z_i is the independent identically distributed (i.i.d) additive complex Gaussian noise with variance σ^2.

For downlink beamforming design, the total transmit power minimisation problem under a given quality-of-service (QoS) constraint is as follows:

$$\underset{w}{\text{maximise}} \quad \sum_{i=1}^{K} \|w_i\|^2$$ (1.10a)

$$\text{s.t.} \quad \text{SINR}_i = \frac{|h_i^H w_i|^2}{\sum_{j \neq i} |h_i^H w_j|^2 + \sigma^2} \geq \gamma_i, \quad \forall i.$$ (1.10b)

where γ_i and SINR_i are the QoS threshold and signal-to-interference-plus-noise ratio (SINR) at the ith user. Note that the problem (1.10) is a non-convex problem since the SINR constraints are non-convex.

To address the problem in (1.10), one approach is to reformulate the beamforming vector in terms of positive semidefinite matrix variables $W_i = w_i w_i^H$ [2]. Thus, the optimisation problem (1.10) is relaxed to the semi-definite programming (SDP:

$$\underset{w}{\text{maximise}} \quad \sum_{i=1}^{K} Tr(W_i)$$ (1.11a)

$$\text{s.t. } Tr \quad (H_i W_i) - \gamma_i \sum_{j \neq i} (H_i W_j) \geq \gamma_i \sigma^2,$$ (1.11b)

$$W_i \succeq 0.$$ (1.11c)

Thus, the problem in (1.11) is a convex problem. In fact, there is always rank-one optimal solution of (1.11), so (1.10) and (1.11) are equivalent.

KKT optimality condition and Lagrangian duality

Karush-Kuhn-Tucker (KKT) condition and Lagrangian duality is the most popular optimality method to solve optimisation problems in wireless communication [4–7]. In this part, we consider the classical KKT optimality conditions which are explicitly analysed for each convex optimisation problem [1, 8]. In general, we consider the following optimisation problem (1.5), which may not necessarily be convex. Let us denote x^* as the local optimal point of problem (1.5) and introduce the Lagrangian multipliers $\lambda \in \mathbb{R}^I$ and $\nu \in \mathbb{R}^J$; the Lagrangian function can be formed as

$$
\begin{aligned}
L(x, \lambda, \nu) &= f_0(x) + \lambda^T g(x) + \nu^T h(x) \\
&= f_0(x) + \sum_{i=1}^{I} \lambda_i g_i(x) + \sum_{j=1}^{J} \nu_j h_j(x).
\end{aligned}
\tag{1.12}
$$

From that, the dual function $p(\lambda, \nu)$ is given as

$$
p(\lambda, \nu) := \min_{x \in \chi} L(x, \lambda, \nu).
\tag{1.13}
$$

Note that $p(\lambda, \nu)$ is always a concave function. Thus, (λ, ν) is a dual feasible if $\lambda \geq 0$ and $p(\lambda, \nu)$ is finite and represented as a lower bound on the primal value $f_0(x) \geq p(\lambda, \nu)$. As a result, $x^* \geq p(\lambda, \nu)$ for all dual feasible vectors (λ, ν).

For a local optimal point (x^*), there exists a unique (λ^*, ν^*) subject to

1. $\nabla_x L(x^*, \lambda^*, \nu^*) = 0$
2. $\lambda_j^* \geq 0$ for $j = 1, ..., I$
3. $\nabla_x L(x^*, \boldsymbol{\lambda}^*, \boldsymbol{\nu}^*) = 0$ for $j = 1, ..., I$
4. $g_j(x^*) \leq 0$ for $j = 1, ..., I$
5. $h_j(x^*) = 0$ for $j = 1, ..., J$
6. $\nabla_{xx} L(x^*, \lambda^*, \nu^*) \geq 0$.

Chapter 2

Recognition and classification of convex programming

There are many benefits of unconstrained and constrained convex optimisation for signal processing and wireless communication. The recognition of a convex optimisation problem should be done before the method for solving the problem by optimisers can be applied. In this section, we will provide some basic approaches to recognising a convex function. A convex optimisation problem is then recognised based on the convexity of its functions.

2.1 Relation to definition

By following the definition of convex function as shown in (1.2), we can recognise a convex function based on this definitions.

2.2 Relation to derivatives

A way to recognise the convexity of functions is based on their differentiability.

2.2.1 First-order conditions

Let χ be a convex set and $f(x)$ be a differentiable function. Then, the $f(x)$ is convex over χ if and only if

$$f(x) \geq f(z) + \nabla_x f(x)(z - x), \ \forall x, z \in \chi, \tag{2.1}$$

where $\nabla_x f(x)$ is the first derivative of the function $f(x)$. Figure 2.1 illustrates this assessment.

The function $f(x)$ is concave on χ if and only if

$$f(x) \leq f(z) + \nabla_x f(x)(z - x), \ \forall x, z \in \chi. \tag{2.2}$$

Some applications of first-order condition are provided in Appendix C.

2.2.2 Second-order conditions

Let χ be a convex set and function $f(x)$ be twice continuously differentiable.

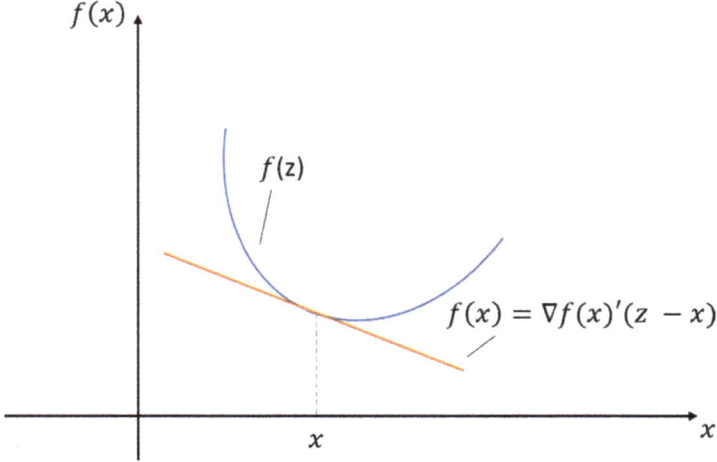

Figure 2.1   Recognising a convex function with differentiability

A twice-differentiable function of single variable is convex if and only if its second derivative is non-negative, i.e.

$$\nabla^2_{xx} f(x) \geq 0, \ \forall x \in \chi,$$ (2.3)

where $\nabla^2_{xx} f(x)$ is the second-order derivative of the $f(x)$.

In contrast, the function $f(x)$ is a concave function on χ if and only if its second-order derivative is non-negative.

Example 2.1: Consider two functions:

$$f_1(x) = \ln(1+x)$$ (2.4a)
$$f_2(x) = \ln(1+1/x)$$ (2.4b)

with $x > 0$. The second-order derivatives of $f_1(x)$ and $f_2(x)$ are

$$\nabla^2_{xx} f_1(x) = -\frac{1}{(1+x)^2} < 0, \forall x > 0,$$ (2.5)

$$\nabla^2_{xx} f_2(x) = \frac{1+2x}{x^2(1+x)^2} > 0, \forall x > 0.$$ (2.6)

Thus, $f_1(x)$ is concave whereas $f_2(x)$ is convex.

Generally, for the variable vector $x \in \mathbb{R}^n$, we need to examine all the partial derivatives of the function $f(x)$. The matrix of all second-order partial derivatives is called the Hessian $H(x)$ of the function $f(x)$.

$$H(x) = \begin{bmatrix} \nabla^2_{x_1 x_1} f(x) & \nabla^2_{x_1 x_2} f(x) & \cdots & \nabla^2_{x_1 x_n} f(x) \\ \nabla^2_{x_2 x_1} f(x) & \nabla^2_{x_2 x_2} f(x) & \cdots & \nabla^2_{x_2 x_n} f(x) \\ \cdots & \cdots & \cdots & \cdots \\ \nabla^2_{x_n x_1} f(x) & \nabla^2_{x_n x_2} f(x) & \cdots & \nabla^2_{x_n x_n} f(x) \end{bmatrix}$$ (2.7)

Theorem 1 *Let f be a function of multiple variables on the convex open set χ with its Hessian. Then,*

1. f *is concave if and only if $H(x)$ is negative semidefinite for all $x \in \chi$.*
2. f *is convex if and only if $H(x)$ is positive semidefinite for all $x \in \chi$.*

Example 2.8: For $x = [x_1, x_2, x_3]^T \in \mathbb{R}^3$, consider the following function:

$$f(x) = x_1^2 + 2x_2^2 + 3x_3^2 + 2x_1x_2 + 2x_1x_3. \tag{2.8}$$

Its Hessian is

$$H(x) = \begin{bmatrix} \nabla^2_{x_1x_1}f(x) & \nabla^2_{x_1x_2}f(x) & \nabla^2_{x_1x_3}f(x) \\ \nabla^2_{x_2x_1}f(x) & \nabla^2_{x_2x_2}f(x) & \nabla^2_{x_2x_3}f(x) \\ \nabla^2_{x_3x_1}f(x) & \nabla^2_{x_2x_3}f(x) & \nabla^2_{x_3x_3}f(x) \end{bmatrix} = \begin{bmatrix} 2 & 2 & 2 \\ 2 & 4 & 0 \\ 2 & 0 & 6 \end{bmatrix}$$

The leading principal minors of the Hessian are $2 > 0, 4 > 0$ and $6 > 0$. Consequently, the Hessian is positive definite and $f(x)$ is convex.

2.3   Relation to convexity propositions

Some useful propositions of the convex/concave function are provided below.

- Non-negative multiple ($\alpha f(.)$):
  $\alpha f(.)$ is convex if $f(.)$ is convex and $\alpha \geq 0$.
- Sum:
  $g(x) = \sum_{i=1}^{m} a_i f_i(x)$ is convex if $f_1, ..., f_m$ are convex and $a_i > 0$.
- Composite:
  $f(Ax + b)$ is convex if $f(.)$ is convex.
- Pointwise maximum:
  $f(x) = \max_{i=1,...,m} \{f_i(x)\}$ is convex if $f_1, ..., f_m$ are convex.
- Piecewise-linear functions:
  $\max_i\{a_i^T x + b_i\}$ where $1 \leq i \leq m$, is convex.
- Quadratic functions:
  $f(x) = x^T Q x + 2q^T x + c$ is convex if $Q \succcurlyeq 0$.
- Piecewise-quadratic functions:
  $\max_i\{x^T Q_i x + q_i^T x + b_i\}$ is convex in x if$Q_i \succcurlyeq 0, 1 \leq i \leq m$.
- Norm functions:
  $\|x\|_k = \left(\sum_{i=1}^{m} |x_i|^k\right)^{\frac{1}{k}}$ where $k \in [1, \infty]$,
- Convex functions over matrix logarithm:
  $-\text{logdet}(X)$ is convex on $X \geq 0$.
- Logarithmic barrier functions:

$$\sum_{i=1}^{m} \log(b_i - a_i^T x)^{-1} \text{ is convex over } P = \{x| a_i^T x \leq b_i, 1 \leq i \leq m\}.$$

Otherwise, we also use Theorem 2 as below.

Theorem 2 *Let C be an open interval. If $f : C \to \mathbb{R}$ is strictly increasing or decreasing then f^{-1} is well defined. In addition, if f is concave or convex, then $f(C)$ is convex and the following holds.*

1. *If f is concave and strictly increasing then f^{-1} is convex.*
2. *If f is concave and strictly decreasing then f^{-1} is concave.*
3. *If f is convex and strictly increasing then f^{-1} is concave.*
4. *If f is convex and strictly decreasing then f^{-1} is convex.*

The proof of Theorem 2 can be found in [1].

2.4   Relation to classes of convex programming

2.4.1   Linear programming

Linear programming (LP) (also called linear optimisation) is a method to achieve the best outcome in a mathematical model whose requirements are represented by linear relationships. From this view, LP is a technique for the optimisation of a linear objective function, subject to (s.t.) linear equality and linear inequality constraints. A general linear problem is given as follows:

$$\underset{x}{\text{minimise}} \quad c^T x + d \tag{2.9a}$$

$$\text{s.t.} \quad Gx \preceq h \tag{2.9b}$$

$$Ax = b \tag{2.9c}$$

where $G \in \mathbb{R}^{m \times n}$, $A \in \mathbb{R}^{p \times n}$ and $c, x \in \mathbb{R}^n$, $\preceq$ is understood componentwise as defined in the Notation.

A modified form of LP is linear fractional programming (LFP) which is also of much interest in engineering. A general LFP problem is to minimise a ratio of affine functions over a polyhedron

$$\underset{x}{\text{minimise}} \quad \frac{c^T x + d}{e^T x + f} \tag{2.10a}$$

$$\text{s.t.} \quad Gx \preceq h \tag{2.10b}$$

$$Ax = b \tag{2.10c}$$

where the domain of the objective function is $\{x | e^T x + f > 0\}$. However, problem (2.10) is not a LP. If the set $\{x | Gx \preceq h, Ax = b, e^T x + f > 0\}$ is feasible, new variables y, z can be introduced to transform (2.10) into an equivalent LP as follows:

$$\underset{x}{\text{minimise}} \quad c^T y + z \tag{2.11a}$$

$$\text{s.t. } Gy - hz \preceq 0 \tag{2.11b}$$

$$Ay - bz = 0 \tag{2.11c}$$

$$e^T y + fz = 1 \tag{2.11d}$$

$$z \geq 0 \tag{2.11e}$$

where $y = x/(e^T x + f)$ and $z = 1/(e^T x + f)$.

Example:

I can buy pizza x_1 and chicken x_2. Pizza costs \$5 per unit and gives me 3 carbohydrates and 2 proteins per unit. Chicken costs \$10 per unit and gives me 5 proteins and 1 carbohydrate per unit. The tasks include:

- maximising the carbohydrates;
- spending no more than $z = 1/(^T + f)$;
- buying at least 10 protein units.

$$\underset{x_1, x_2}{\text{maximise}} \quad 3x_1 + 1x_2 \tag{2.12a}$$

$$\text{s.t.} \quad 5x_1 + 10x_2 \leq 30 \tag{2.12b}$$

$$2x_1 + 5x_2 \geq 10 \tag{2.12c}$$

$$x_1, x_2 \geq 0, \text{integer} \tag{2.12d}$$

2.4.2 *Quadratic programming*

An optimisation problem is called convex quadratic program (CQP) if the objective function is convex quadratic and the constraints are linear.:

$$\underset{x}{\text{minimise}} \quad \frac{1}{2}x^T P x + q^T x + r \tag{2.13a}$$

$$\text{s.t. } Gx \preceq h \tag{2.13b}$$

$$Ax = b \tag{2.13c}$$

where $P \in S_+^n$, $G \in \mathbb{R}^{m \times n}$, $A \in \mathbb{R}^{m \times n}$ and $q, x \in \mathbb{R}^n$.

Example: Consider CQP [9]

$$\text{minimise} \quad f(x) = \frac{1}{2}x^T \begin{bmatrix} 4 & 0 & 0 \\ 0 & 1 & -1 \\ 0 & -1 & 1 \end{bmatrix} x + \begin{bmatrix} -8 \\ -6 \\ -6 \end{bmatrix} x + r \tag{2.14a}$$

$$\text{s.t.} \quad x_1 + x_2 + x_3 = 3 \tag{2.14b}$$

$$x \geq 0. \tag{2.14c}$$

The problem can be solved by optimisation tool packages (i.e. CVX). Its solution found as

$$x^* = \begin{bmatrix} 0.5 \\ 1.25 \\ 1.25 \end{bmatrix}$$

2.4.3   Second-order cone programming

Close to QP, a problem of second-order cone program (SOCP) has a common form as

$$\text{minimise} \quad c^T x \qquad \qquad (2.15a)$$
$$\text{s.t.} \quad \|A_i x + b_i\|_2 \le d_i^T x + e_i, i = 1, ..., I \qquad (2.15b)$$
$$Fx = g, \qquad \qquad (2.15c)$$

where $A_i \in \mathbb{R}^{m_i \times n}$, $b_i \in \mathbb{R}^{m_i}$, $F \in \mathbb{R}^{p \times n}$, $g \in \mathbb{R}^p$ and $c, d_i, x \in \mathbb{R}^n$.

A constraint in the form of

$$\|Ax + b\|_2 \le d^T x + e \qquad \qquad (2.16)$$

is called a second-order cone constraint.

Example: Consider the problem below:

$$\text{minimise} \quad \sum_{i=1}^{I} \frac{1}{a_i^T x + b_i} \qquad \qquad (2.17a)$$
$$\text{s.t.} \quad a_i^T x + b_i > 0, \ i = 1, ..., I, \qquad (2.17b)$$
$$d_j^T x + e_j \ge 0, j = 1, ..., J, \qquad (2.17c)$$

which is convex since the objective function (1.17a) is convex. This problem can be expressed as SOCP by introducing new variables t_i and writing the problem as a hyperbolic constrained problem.

$$\text{minimise} \quad \sum_{i=1}^{I} t_i \qquad \qquad (2.18a)$$
$$\text{s.t.} \quad t_i(a_i^T x + b_i) \ge 1, t_i \ge 0, i = 1, ..., I, \qquad (2.18b)$$
$$d_j^T x + e_j \ge 0, j = 1, ..., J. \qquad (2.18c)$$

Many types of convex problems can be transformed as SOCPs by using hyperbolic constraints as below

$$x \ge 0, \ y \ge 0, \ t^2 \le xy \Leftrightarrow \left\| \begin{bmatrix} 2t \\ x-y \end{bmatrix} \right\| \le x + y. \qquad (2.19)$$

Following that, (2.18) can be cast as SOCP

$$\text{minimise} \quad \sum_{i=1}^{I} t_i \qquad \qquad (2.20a)$$
$$\text{s.t.} \quad \left\| \begin{bmatrix} 2 \\ a_i^T x + b_i - t_i \end{bmatrix} \right\| \le a_i^T x + b_i + t_i, i = 1, ..., I, \qquad (2.20b)$$
$$d_j^T x + e_j \ge 0, j = 1, ..., J. \qquad (2.20c)$$

2.4.4 Geometric programming

A geometric program (GP) is formed as

$$\underset{x}{\text{minimise}} \quad f_0(x) \tag{2.21a}$$

$$\text{s.t.} \quad f_i(x) \le 1, i = 1, ..., I \tag{2.21b}$$

$$h_j(x) = 1, j = 1, ..., J \tag{2.21c}$$

where $f_0, f_1, ..., f_I$ are posynomials and $c > 0$ are monomials.

A monomial function is defined as $h : \mathbb{R}^n \to \mathbb{R}$ with $\text{dom} f = \mathbb{R}^n_{++}$

$$h(x) = cx_1^{a_1} x_2^{a_2} ... x_n^{a_n}, \tag{2.22}$$

where $c > 0$, $a_i \in \mathbb{R}$.

A posynomial is a sum of monomial functions such as

$$f(x) = \sum_{k=1}^{K} c_k x_1^{a_{1k}} x_2^{a_{2k}} ... x_n^{a_{nk}}, \tag{2.23}$$

where $c_k > 0$, $a_{ik} \in \mathbb{R}$.

The general form of GP may not be a convex program. By change the variables and transformation techniques of the objective and constraint functions, some GP problems can be transformed into convex optimisation problems.

Following the approach in [1], the change of variables is made as $y_i = \log x_i$ so $x_i = e^{y_i}$. The function $h(x)$ is reformed as

$$\begin{aligned} h(x) &= h(e^{y_1}, ..., e^{y_n}) \\ &= c(e^{y_1})^{a_1} ... (e^{y_n})^{a_n} \\ &= e^{a^T y + b} \end{aligned} \tag{2.24}$$

where $a = [a_1, ..., a_n]^T$, $y = [y_1, ..., y_n]^T$ and $b = \log c$. The new variables y_i turn the monomial function $h(x)$ into an exponential-affine form.

Similarly, the posynomial function $f(x)$ is reformed as

$$f(x) = f(e^{y_1}, ..., e^{y_n}) = \sum_{k=1}^{K} e^{a_k^T + b_k} \tag{2.25}$$

where $a_k = [a_{1k}, ..., a_{nk}]^T$ and $b_k = \log c_k$.

By taking the logarithm, the GP (2.21) can be expressed in terms of the new variable y as

$$\underset{y}{\text{minimise}} \quad \tilde{f}_0(y) = \log \left(\sum_{k=1}^{K} e^{a_{0k}^T y + b_{0k}} \right) \tag{2.26a}$$

$$\text{s.t.} \quad \tilde{f}_i(y) = \log \left(\sum_{k=1}^{K} e^{a_{ik}^T y + b_{ik}} \right) \le 0, \forall i \tag{2.26b}$$

$$\tilde{h}_j(y) = g_j^T y + h_j = 0, \forall j \tag{2.26c}$$

where $\tilde{f}_i$ is convex and $\tilde{h}_j$ is affine. Thus, the problem (2.26) is a convex optimisation problem.

Example: Let us consider the problem below:

$$\underset{x,y,z}{\text{minimise}} \quad x^{-1}y^{-1/2}z^{-1} + 2.3xz + 4xyz \tag{2.27a}$$

$$\text{s.t.} \quad \frac{1}{3}x^{-2}y^{-2} + \frac{4}{3}y^{1/2}z^{-1} \leq 1 \tag{2.27b}$$

$$x + 2y + 3z \leq 1 \tag{2.27c}$$

$$\frac{1}{2}xy = 1 \tag{2.27d}$$

This is a standard form of GP, with three variables, two inequality constraints and one equality constraint. This problem can be solved by using an optimisation tool package with a geometric programming solver.

2.4.5 Semidefinite programming

We consider a general form of semi-definite programming (SDP) in the form of minimisation of a linear function of a vector variable $x \in \mathbb{R}^M$ [10]:

$$\underset{x}{\text{minimise}} \quad c^T x \tag{2.28a}$$

$$\text{s.t.} \quad F(x) \geq 0, \tag{2.28b}$$

where (2.28b) represents a matrix inequality defined as

$$F(x) = F_0 + \sum_{m=1}^{M} x_m F_m$$

for $F_0, F_1, ..., F_M \in \mathbb{R}^{N \times N}$. The inequality (2.28b) results in $F(x) \geq 0$, indicating that $F(x)$ is positive semidefinite. A positive semidefinite function has to satisfy the condition that $z^T F(x)z \geq 0$ for all $z \in \mathbb{R}^N$. The inequality (2.28b) is also called a linear matrix inequality (LMI).

SDP has been adopted to solve many engineering design problems since it unifies several standard problems including LP and QP. Furthermore, SDP is not much harder to solve by using interior-point methods. Therefore, SDP is one of the most general optimisation programs [10].

Example: Consider the following non-linear optimisation problem:

$$\underset{x}{\text{minimise}} \quad \frac{(c^T x)^2}{d^T x} \tag{2.29a}$$

$$\text{s.t.} \quad Ax + b \geq 0, \tag{2.29b}$$

where $d^T x > 0$.

An approach to solving problem (2.29) is by introducing an auxiliary variable t. The added variable serves as an upper bound on the objective function

$$\underset{x}{\text{minimise}} \quad t \tag{2.30a}$$

$$\text{s.t.} \quad Ax + b \geq 0, \tag{2.30b}$$

$$t \leq \frac{(c^T x)^2}{d^T x} \quad . \tag{2.30c}$$

Then, we can reformulate problem (2.30) as a SDP problem by expressing a LMI in x and t as:

$$\underset{x}{\text{minimise}} \quad t \tag{2.31a}$$

$$\text{s.t.} \quad \begin{bmatrix} \text{diag}(Ax + b) & 0 & 0 \\ 0 & t & c^T x \\ 0 & c^T x & d^T x \end{bmatrix} \quad . \tag{2.31b}$$

This SDP problem can be solved by using an optimisation tool package with a SDP solver.

2.5   Relation to equality and inequality

Please see more details in Appendix B and C.

Chapter 3

Convex optimisation for signal processing and wireless communication

Optimisation is often applied to solve a wide range of problems in communications and signal processing such as system design, filter design, resource allocation, antenna design or, in general, any task that involves convex optimisation concepts. In each design problem, optimisation methods should be recognised and appropriate techniques should be selected to handle the problem. Convex optimisation has played an important role in model analysis, algorithm design and network performance optimisation [8,11]. Since there are numerous scenarios of convex optimisation in wireless communication, let us briefly introduce some specific aspects of wireless communication applications that use optimisation in recent years.

3.1 Convex optimisation for signal estimation

In this section, we look into some deterministic and stochastic approaches to signal estimation for signal processing such as matched filter and Wiener filter. The former is used in the case of a single signal in noise to maximise the output signal-to-noise ratio whereas the latter is used in the case of several sources in noise to maximise the output signal-to-interference plus noise (SINR). These filters are efficient if the channel system is estimated well or the segment of transmitted data is known based on a training sequence. The improvement of filter accuracy is needed by using spatial processing, parametric data updating and beamforming techniques.

Example: We consider a simple array signal processing model as follows:

$$x(t) = \sum_{i=1}^{L} a_i s_i(t) + n(t) = As(t) + n(t), \tag{3.1}$$

The where the signal is sampled at integer time instants, i.e., is noise model. Then, we have

$$x_k = x(k) = \sum_{i=1}^{L} a_i s_i(k) + n(k) = \sum_{i=1}^{L} a_i s_{i,k} + n_k = As_k + n_k, \tag{3.2}$$

where $x_k \in \mathbb{C}^N$ is the received signal from the transmitter equipped with N antennas. The elements of signal vector s_k are zero mean and independent and, thus, uncorrelated $\mathbb{E}[s_k s_k^H] = \mathbf{I}$. We assume that the system collects M sample vectors. The matrix $\mathbf{X} \in \mathbb{C}^{N \times M}$ can be obtained as

$$X = AS + N, \tag{3.3}$$

where $S \in \mathbb{C}^{L \times M}$ contains all the source signal samples.

Now, we first consider the *noiseless case*. Our data process is given by:

$$X = AS. \tag{3.4}$$

The main work is to design a linear beamforming matrix **W** such that

- **A** is known, then $W^H = A^\dagger, S = W^H X$
- **S** is known, then $W^H = SX^\dagger, A = (W^H)^\dagger$

where $A^\dagger = (A^H A)^{-1} A^H, (L \leq N)$ is the Moore-Penrose pseudo-inverse of A, $X^\dagger$ is a right inverse of **X**. In both cases, the beamformer is exactly found by cancelling out all interference and noise.

In the second case, we consider the presence of additive noise in the system, *noise case*, i.e., $X = AS + N$. We use linear least squares method for minimising the model fitting error or the output error. The former approach is shown as

$$\min_{S} \|X - AS\|_F^2, \tag{3.5}$$

with **A** being known, or

$$\min_{A} \|X - AS\|_F^2, \tag{3.6}$$

with **S** being known.

The latter approach is based on minimising the output error as

$$\min_{W} \|W^H X - S\|_F^2, \tag{3.7}$$

with **A** or **S** being known.

3.2   Convex optimisation for resource allocation problems

Many resource allocation problems of wireless communication are considered as optimisation problems [12–15], for example, power minimisation or allocation problem under quality-of-service (QoS) constraints and depleting energy resources [16,17], reverse problem of the SINR or sum rate maximisation [18–21]. In other tasks, interference that always exists in wireless communication models has to be minimised in order to enhance desired signals [22–24]. The combination of a beamforming design and power minimisation or sum-rate maximisation scheme is an efficient way to significantly improve the network performance while handling the interference terms caused by users of neighbouring cells (inter-cell interference) or other users within the same cell (intra-cell interference) [16,24–27].

Example: In [13], the authors considered an orthogonal frequency-division multiplexing system with K users, N subchannels and a power budget P_{total}. Based on joint subchannel and power allocation, the weighted sum-rate maximisation problem is shown as

$$\max_{p_{k,n}, \rho_{k,n}} \sum_{k=1}^{K} R_k = \sum_{k=1}^{K} \sum_{n=1}^{N} \frac{\rho_{k,n}}{N} \log_2(1 + \frac{p_{k,n} h_{k,n}^2}{N_0 B/N}) \tag{3.8a}$$

$$\text{s.t.} \sum_{k=1}^{K} \sum_{n=1}^{N} p_{k,n} \leq P_{\text{total}} \tag{3.8b}$$

$$p_{k,n} \geq 0, \forall k, n \tag{3.8c}$$

$$\rho_{k,n} = \{0, 1\}, \forall k, n \tag{3.8d}$$

$$\sum_{k=1}^{K} \rho_{k,n} \leq 1, \forall n \tag{3.8e}$$

$$R_1 : R_2 : \dots : R_K = \gamma_1 : \gamma_2 : \dots : \gamma_K, \tag{3.8f}$$

where B and N_0 are available bandwidth and power spectral density of additive white Gaussian noise channel; $h_{k,n}$ and $p_{k,n}$ are the channel gain and the power allocated to user k in subchannel n. Equations (3.8b) and (3.8c) represent the power constraints. $\rho_{k,n}$ provided in (3.8d) indicates whether subchannel n is used by user k or not. Constraint (3.8e) shows that each subchannel can only be used by one user.

The above optimisation problem is very difficult to handle as it is a nonconvex one. To this end, the authors in [13] derived a suboptimal algorithm to deliver a performance solution close to the global optimum. The proposed approach is to transform the problem into a convex one by relaxing nonconvex constraints or applying linearisation procedure. More details can be found in [13].

3.3 Convex optimisation for the problems of scheduling and deployment in wireless networks

Densification, such as adding more base stations (BSs) and access points, is one of the most important strategies that enable throughput increases in wireless networks. This is set to continue in 5G and beyond. However, this approach leads to many challenges. It can cause strong aggregate interference issues as a consequence of more spatial spectrum reuse. Furthermore, effective power consumption schemes should be in place in order to save the amount of energy consumed. For example, in heterogeneous networks (HetNets), a large number of BSs consume a huge amount of energy during their operation in the long run. A typical universal mobile telecommunications service BS consumes 800–1500W at a radio frequency output power of 20–40W, and they still reach more than 90% of their overall power consumption even with little and no activity.

Convex optimisation methods have various applications in the scheduling and deployment of wireless communication systems [28,29]. Novel transmission strategies using optimisation techniques are also efficient for providing high performance in wireless networks densification [29–35].

Example: In [34], the authors propose a linear transmit beamforming method to solve a joint utility maximisation and regularised weighted mean square error problem. The optimisation problem is formulated as

$$\max_{V_{i_k}^{q_k}} \sum_{k \in \mathcal{K}} \sum_{i_k \in \mathcal{I}_k} \left(u_{i_k}(R_{i_k}) - \lambda_k \sum_{q_k \in \mathcal{Q}_k} \| V_{i_k}^{q_k} \| \right) \tag{3.9a}$$

$$\text{s.t.} \sum_{i_k \in \mathcal{I}_k} (V_{i_k}^{q_k})^H V_{i_k}^{q_k} \le P_{\text{total}}^{q_k} \tag{3.9b}$$

where $u_{i_k}(.)$ represents the utility function of the ith user, λ_k is the control parameters for sparsity level in each cell and R_{i_k} is the achievable rate for user i_k. The induced group-sparse structure of the beamformers $\{V_{i_k}\}_{i_k \in \mathcal{I}_k}$ is proposed for optimising the system performance. As the utility functions and beamforming sparsity terms in objective (3.9a) are non-deterministic polynomial-time hardness (NP-hardness), the above optimisation problem is difficult to solve. To handle this, the authors transformed the problem such that the reformulated problem is a series of convex problems, and thus, a stationary solution can be found.

3.4   Convex optimisation for emerging wireless network technologies

Since wireless networks depend on the environment and geometric modelling in each area, there is no general model or technology for any network. Instead, researchers need to study and develop an appropriate technology for each type of wireless communication system [36–38]. HetNets [39–42], multiple-input multiple-output (MIMO) and massive MIMO [43–45], cognitive radio networks [46–48], cell-free networks [49–51], millimetre-wave networks [52–54], wireless energy harvesting [55–58] and physical layer security [59–62] are some of the emerging wireless models and technologies aiming at providing wireless services to all network users at a good QoS, and ubiquitous and high data rate connectivity. Along with their envisioned benefits, these emerging networks bring numerous challenges such as the allocation and management of radio spectrum, co-existence of different networks, explosively increased energy consumption or interference issues. The effective deployment of these networks relies on their optimal modelling and design as well as on optimisation method and algorithms for the optimal management and utilisation of radio spectrum and consumption of energy resources.

Example: In our work [51], we considered a cell-free massive MIMO system with bandwidth B. K single-antenna users are served by M randomly deployed single-antenna access points in the same time-frequency resource. When it comes to exploiting a massive number of access points in a large-scale network, energy efficiency (EE) performance is a major figure-of-merit. It is critical to meet the ratio of transmitted bits and total power consumption in green communications, while satisfying QoS of the network. The EE maximisation problem is formulated as:

$$\max_{\eta} \frac{B \cdot r(\eta)}{P_{\text{total}}(\eta)} \tag{3.10a}$$

$$\text{s.t.} \ r_k(\eta) \ge \bar{r}_k, \ k = 1, \dots, K, \tag{3.10b}$$

$$\sum_{k=1}^{K} \theta_{mk}\eta_k \leq 1, m = 1, ..., M, \tag{3.10c}$$

$$\eta_k \geq 0, k = 1, ..., K. \tag{3.10d}$$

Therein,

$$r(\boldsymbol{\eta}) = \sum_{k=1}^{K} r_k(\boldsymbol{\eta}) = \sum_{k=1}^{K} (1 - \frac{\tau_u}{\tau}) \log_2 \left(1 + \frac{\rho_f \eta_k}{1 + \rho_f \sum_{i=1}^{K} \gamma_{ki} \eta_i} \right) \tag{3.11}$$

is the total spectral efficiency of the system using zero-forcing precoding, and

$$P_{\text{total}} = P_{\text{cir}} + \sum_{m=1}^{M} P_m + \sum_{m=1}^{M} P_{\text{bh},m}, \tag{3.12}$$

is the total power for the downlink transmission.

In problem (3.10), constraint (3.10b) represents the QoS requirement for each user. Equations (3.10c) and (3.10d) represent the power control constraints. If the channel estimation is assumed perfect, problem (3.10) belongs to the class of convex optimisation problems as a fractional programming and, thus, can be solved by Dinkelbach's algorithm [63].

3.5 Convex optimisation for smart wireless networks

State-of-the-art wireless communication, such as 5G and beyond, will see an enhancement in many aspects, for example, spectral efficiency, EE, low latency, high reliability and real-time applications [64–66]. Green communication is one of the most recent trends which will require a lot of attention in terms of optimisation applications [67–69]. Security is another research hotspot, given there are emerging ways of jamming or eavesdropping the desired signal of the link from source to destination [59].

Example: In our work [25], we considered a two-tier downlink network, in which one macro BS and S small-cell BSs share the same frequency spectrum as illustrated in Figure 3.1. All BSs, which are connected to a central processor via backhaul links, are supposed to cooperate to serve K users, each of which is equipped with a single antenna.

A three-objective (EE, QoS and service loading) optimisation problem is proposed as

$$\max_{\mathbf{T},t} f(\mathbf{T}, t) - \gamma g(\mathbf{T}) \tag{3.13a}$$

$$\text{s.t. } \ln(1 + |\mathbf{h}_k \mathbf{t}_k|^2/\sigma_k^2) \geq \bar{C}_k, \forall k, \tag{3.13b}$$

$$\sum_{k=1}^{K} \|\mathbf{G}_k^s \mathbf{t}_k\|^2 \leq P_{\text{max}}^s \tag{3.13c}$$

$$\sum_{k=1}^{K} \left[\mathbf{G}_k^s \mathbf{t}_k (\mathbf{G}_k^s \mathbf{t}_k)^H \right]_{\ell,\ell} \leq P_{\ell,\text{max}}^s \tag{3.13d}$$

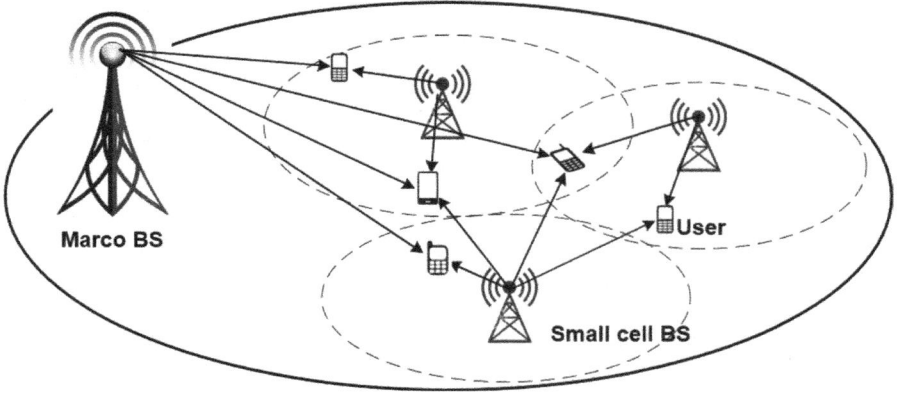

Figure 3.1   An example model for downlink HetNets

$$\sum_{s=0}^{S} \frac{1}{\lambda_s} \sum_{k=1}^{K} \|\mathbf{G}_k^s \mathbf{t}_k\|^2 + P_{\text{cir}} \leq t \tag{3.13e}$$

where $\gamma > 0$ is the sparsity penalty parameter. Therein,

$$f(\mathbf{T}, t) = \frac{\sum_{k=1}^{K} \ln(1 + |\mathbf{h}_k \mathbf{t}_k|^2 / \sigma_k^2)}{t}, \tag{3.14}$$

$$g(\mathbf{T}) = \sum_{(k,s) \in \mathcal{K} \times (\mathcal{S} \setminus \{0\})} ln(1 + \left\| \mathbf{G}_k^s \mathbf{t}_k \right\|^{2/\epsilon}) \tag{3.15}$$

The inclusion of the nonconvex functions in (3.13) makes the problem more computationally challenging. A novel path-following computational procedure is proposed for solving the problem, which invokes a simple convex quadratic program of moderate size at each iteration and converges to at least a locally optimal solution. The reader can see the detailed solution in [25].

Chapter 4

Introduction to real-time embedded optimisation programming

In this chapter, we review major aspects of real-time embedded optimisation programming, such as the concept and program structure of real-time operation, timing complexity analysis and specification frameworks for real-time embedded systems. Our analysis gives a comprehensive overview of real-time optimisation programming and demonstrates the effectiveness and applicability of the real-time optimisation approach for the design of engineering systems.

4.1 Concepts of real-time systems

One widely adopted definition of real-time systems describes them as real systems that are capable of meeting the requirements of real-world problems within a limited processing time. In other words, the implementation of real-time applications will have to satisfy timing constraints, e.g., solving deadlines and other timing requirements. This leads to the development of real-time programs in both software and hardware systems based on timing properties and characteristics. The most important features of time-critical systems are effective strategies, robust algorithms, parallelism techniques and final processing speed. The question is, how processing speed corresponds to timing demands and program statements. A carefully analysed problem statement and resource allocation of the system are tremendously needed to approach timing demands. In general, scheduling policies are useful in allocating the system's essential resources to embedded programming. By calculating the precision of development stages, schedulers can be determined. Then, specifications, verification and timing analysis will be collaboratively performed to provide an accurate prediction of the timing in real-time systems.

With respect to external program events, there are two categories of real-time scenarios in most time-critical systems, namely synchronous real-time and asynchronous real-time scenarios [70]. In synchronous real-time systems, the external program events are assumedly ordered in time and instantaneously responsive to each stage event. These scenarios are suitable for distributed systems implementing a single processor. In asynchronous real-time cases, external events occur at a dense time domain under a response-time bound. Some restrictions need to be imposed on the system in order to analyse timing behaviour. These approaches enable full determination of timing properties for asynchronous models, i.e., imposing determinism, approximating cyclic behaviour and periodic behaviour. An asynchronous scenario can be the most popular in real-time systems since it can be applied in single-processor, multi-processor and distributed systems.

Based on the perspectives of real-time systems, we can classify real-time systems into hard versus soft real-time, fail-safe versus fail-operational, guaranteed-timeliness versus best-effort, resource-adequate versus resource-inadequate and event-triggered versus time-triggered [71]. Below we discuss the fundamental characteristics of hard and soft real-time systems.

- Response time:
  - Hard real-time systems are used to support response time of milliseconds or less. Human intervention often is very restricted. Hence, hard real-time systems highly require autonomy to maintain process operations.
  - Soft real-time systems require response time in the order of seconds and more. Even when the time deadline is broken, there are still no serious threads on the system.
- Peak-load performance:
  - In hard real-time systems, peak-load performance needs to be well-defined in order to guarantee that strict deadlines are met in all program stages.
  - In soft real-time systems, the average-load performance is often more focused on processing stages. For the simplicity of system design or economical considerations, the degraded performance of operation caused by peak-load cases can be accepted.
- Size of data files:
  - Hard real-time systems concern the accuracy in short-term of real-time database which is constituted by small data files.
  - Soft real-time systems concern the maintenance of large data files for long-term integrity since they may not require a low latency or rapid response time.
- Redundancy methods:
  - Hard real-time systems require very fast operational computing. Roll-back or recovery actions are often limited in these systems because of the difficulties in guaranteeing deadlines and the difference in checkpoint time.
  - Roll-back/recovery methods can be easily applied in soft real-time systems after operational errors are detected.

In terms of real-time support, real-time systems can be categorised into guaranteed-response and best-effort designs. In guaranteed-timeliness systems, taking into account the peak load and the number and type of faults in practice, a guaranteed response must be carefully planned and extensively analysed under a system failure probability. In best-effort systems, a 'best possible effort taken' can be designed when a guaranteed response cannot be given. The latter approach does not require a rigorous technique to guarantee the system failure probability, and, hence, these systems are suitable for fast implementation in critical real-time scenarios without high-precision pointing performance requirements.

4.1.1   Modelling real-time systems

In a real-time system, there are many algorithms and programs for data transformation and processing to compute the desired results. These programs can be described on a level of abstraction that considers the following aspects in the data domain and time domain [71]:

- input data given
- internal state of the program
- intended results
- modifications to the internal state of the program and
- resource requirements of the program, e.g. memory size, speed processing.

In the time domain, the execution time is very important to derive the computing results by initiating the first computation step from the input data and control signal to the final computation step. Depending on the structural elements of real-time operating systems, a set of computational clusters can be formed by a decomposition process from a distributed real-time application. Otherwise, a computational cluster can be further partitioned into a set of fault-tolerant units (FTUs) connected by a real-time local area network. A set of concurrently executing tasks performs intended functions within a node computer in each FTU.

- A computational cluster is to perform the intended fault-tolerant service for the cluster environment. The cluster environment consists of the control object, the operator and other computational clusters. Gateway nodes of the cluster generate the interfaces between a cluster and its environment to enable interconnected computational clusters in the form of a mesh network.
- A computer node is a self-contained computer including both hardware and software and performs a set of well-defined functions. A node will bind software and hardware resources into a single operational unit with observable behaviour for both temporal and value domains in a distributed real-time system.
- A task is the execution of a sequential program starting with processing input data and internal state, and terminating with producing results and updating the internal state. The time interval between the start and termination in each task is called the actual duration of the task on a given target machine.

4.1.2   Real-time dynamic scheduling

Dynamic scheduling of real-time systems requires a sequence of decisions that are taken by assigning system resources including processors, memory and shared data structures, to tasks. The tasks then can provide arbitrary attributes, e.g., arrival times, resource requirements, computational times, deadlines and other important values. Dynamic algorithms are needed for applications where computing requirements may vary continuously and widely, making fixed priority scheduling difficult or inefficient. Real-time applications such as these in robotics and time-critical mission services essentially require support from dynamic scheduling. In such applications, the control system must adapt to a dynamic environment to allow for more flexibility in dealing with practical issues, such as the need to alter scheduling decisions based on the occurrence of overloads.

Dynamic scheduling involves three basic steps including feasibility checking, schedule construction and dispatching. Feasibility or schedulability analysis is a process determining whether the timing requirements of a set of tasks can be satisfied under a given set of resource requirements and precedence constraints. When tasks arrive, dynamic systems perform feasibility checking online. The programming model can be adopted, and the scheduling algorithm is provided and used depending on the designed application in the system.

Here, we discuss two approaches to dynamic scheduling in real-time systems [71]:

1. Planning-based approach: The execution of a task will only begin if a feasibility test shows that it is always completed before its deadline. A schedule or plan is used to decide when a task should begin execution. Before the execution of a set of tasks, the feasibility of capturing them is checked under scheduling policies such as 'earliest-deadline-first' or 'least-laxity-first'.
2. Best-effort approach: There is no need for feasibility checking before the tasks are queued; the system tries to 'do its best' to meet deadlines; however, a task may be aborted during its execution. Tasks may be queued according to scheduling policies that take account of the time constraints.

Task completion is important to scheduling decision-making regardless of whether or not feasibility checking is done. This observation is proposed as a time-value function that identifies the task contributions for the system upon its successful completion. The time-value function quickly drops 'hard' real-time tasks after the deadline and, thus, dynamic algorithms cannot be used. Dynamic algorithms are appropriate for the tasks in the categories of 'soft' real-time tasks.

The relative values of tasks, i.e., rejected tasks and executed tasks, are considered in determining system performance. In dynamic scheduling algorithms, the total value is not easily predicted since the prior knowledge of the tasks is unknown. The algorithm must attempt to maximise the accrued value from the previous tasks that have been completed on time. On the other hand, for most dynamic algorithms, an assignment process can be developed, in which a value function is assigned a positive value to successfully completed a task and zero to an incomplete task. In real-time systems, the capability for 'graceful degradation' is exhibited while the achieved value is optimised. To this end, not only any missed deadline must be detected after a task has been executed, but the fact is that this should be detected as soon as possible. Early detection is needed to make it possible for the task to be substituted by one or more contingency tasks. Hence, real-time scheduling analysis must have an early warning feature that provides sufficient lead time for the timely invocation of contingency tasks, corresponding to the continuously changing environment.

For complex error situations, a transaction is enabled to take place only if it can guarantee the completion by its deadline. In the opposite case, the program can perform an alternative action under an imposed deadline by the determination of schedulability. This process can be generalised to multiple transactions as N versions and the best possible version can be executed by the guarantee of algorithms used.

4.1.3 Real-time communication

This section investigates the requirement of low-latency processing in real-time systems. Tasks cooperate with one another to exchange their information and then coordinate their activities together for an embedded multi-thread application. This represents signal-centric, data-centric approach or both. The necessary information is exchanged within the event signal itself in signal-centric communication, meanwhile, the information is carried within the transferred data in data-centric communication. Lastly, data transfer accompanies event notification in their combination. To present the communication in real-time systems, we discuss basic required materials as below:

- Protocol latency
  - Protocol latency defines a time interval between the delivery of message transmission between the transmitting node and the receiving node. To guarantee a consistent behaviour of the system, the message should be permanently delivered across a communication network interface to the hosting node.
- Flexibility
  - Real-time communication systems must adapt and support different system configurations over time. A real-time protocol should be flexible to accommodate the changes without requiring a software modification or retesting operational nodes. However, with the limited bandwidth of the communication channel, the increase in communication traffic must be controlled within given time constraints.
- Error detection
  - The effective communication system must provide predictable and dependable services, even error detection. When errors have occurred during the message transmission, they must be immediately detected and handled with a very small jitter. If an error cannot be addressed, all the communicating partners will be informed about that situation together with an alternative solution.
- Synchronisation
  - Synchronisation is classified into resource synchronisation and activity synchronisation. The former identifies the authorised access to a shared resource. The latter determines the authorised execution of a multi-threaded program.
  - An application's design typically involves multiple concurrent threads, tasks or processes. In real-time embedded systems, applications can use

both synchronisation types to maximise system efficiency. Coordinating the activities for application design purposes requires inter-task synchronisation and communication and ready-to-use embedded design patterns.

4.1.4   *Real-time performance analysis*

As performance analysis can be implemented in all stages of the system operation life cycle, a predictive performance analysis process is necessary for design and programming phases in a real-time system. Often, the performance of critical algorithms, achievable response times and task schedulability are analysed before the complete system operation. Specific performance measures will be estimated using collective knowledge from software products, system-level simulations and so forth. In the testing phase, it is also practical to measure the performance in a real operating environment.

Simplified estimation approaches such as a handy toolset should be used in real-time performance analysis with short response times. In computational complexity theory, the complexity class P (polynomial) is the class of problems that can be solved by an algorithm that runs in polynomial time on a deterministic computing machine. On the other hand, the complexity class NP (non-polynomial) is the class of all problems that cannot be solved in polynomial time. If a problem belongs to the class NP, particular decision-making or problem recognition is considered to be a NP-complete task.

The challenges in finding optimal solutions for the problems of real-time scheduling have been a potential research direction. However, real-time scheduling often requires practical constraints to be managed or classified as NP-complete [72]. Below are representative features of the performance analysis in real-time systems:

- When there are mutually exclusive constraints, it is impossible to find a totally online optimal run-time scheduler.
- The problem of deciding whether it is possible to schedule a set of periodic tasks is used to enforce mutual exclusion. Often, these tasks are NP-hard.
- A multiprocessor scheduling problem with two processors, no resources, arbitrary partial-order relations and every task having a one-unit computation time is polynomial.
- A scheduling problem with multiple processors, no resources, independent tasks and task computation times is NP-complete.
- A scheduling problem with multiple processors, one resource, all independent tasks and the computation time of every task equal to 1 unit is NP-complete.
- In multi-processor systems, earliest deadline scheduling is not efficient and optimal.
- For multi-processor operation, deadline scheduling algorithms can only be optimised with complete *a priori* knowledge of deadlines, computation times and task start times.

Real-time performance analysis is a priority as it is to check if the system will meet critical deadlines. However, this task is rarely possible due to the NP-completeness of scheduling problems and severe constraints of synchronisation mechanisms in practice. Fortunately, an approximate analysis is sufficient and feasible to handle the system behaviour. Prediction, estimation and measurement of the execution time at essential code units are important steps in performing schedulability analysis in real-time systems.

Before implementation, an estimation of the execution time for certain program modules and the overall system operating time is necessary from both engineering and project management perspectives. Requirements of CPU utilisation should be planned and expressed for specific design goals, e.g., the selection of embedded processor platform for real-time applications. This step aims at identifying problematic code units such as slow or inadequate response times during programming and testing phases. Logic-analyser approach is one method in real-time performance analysis where all hardware latencies, system delays and uncertainties are taken into account. Hence, logic analyser is usually deployed in the final stages of programming, during testing and system integration.

In addition, the determination of instruction execution times should also be taken into consideration in real-time performance analysis. This measurement depends on memory access times and waiting states at the source region of the instruction code. Some real-time systems of specific processing platforms usually use simulation software to estimate instruction execution times and CPU throughput. Accordingly, users can select CPU type, memory speeds, instruction mix and calculate total instruction times and throughput for different design goals.

4.2   Real-time computing

A system can respond to real-time computing if its ability to process tasks satisfies real-time-bound responses for performing necessary computations. Physical components are controlled by functions of a real-time computing framework and then they process tasks at periodic intervals and respond by sending signals to the main system controller and other components [70]. A time bound is required for all types of delivered responses. In time-critical missions, a computation schedule is essential to support the required time bounds for each response in a sequence of events. In theory, it may be unable to meet timing constraints within the limited processing capacity available. When the time bound of a response or the timing constraint is not met, there may be a catastrophic system failure. Therefore, any real-time program has to be considered and designed carefully on a real system with finite resources. Their computing platform should be adapted to a time-varying environment, interactive with predicted time-dependent behaviours and executed with time-bound responses.

However, unlike the human brain's ability to react in time, the assessment of real-time demands on a computer system is a difficult task to implement and, thus, how do we determine whether they will be met? [70]. When the number of tasks to perform with large processing loads increases, multiple processors as a potential technology for multiple tasks in real-time processing are required. To this end, the schedule of multi-processing needs to be designed to interact with different conditions of observed tasks, predict and synchronise with time-bound responses for more than one processor at the same time and flexibly distribute system resources at any time to calculate accurately. Last, to complete the dependent tasks, the processing system on a multi-processing platform has to ensure timely information exchange between tasks – this can be solved by setting up some time-bound constraints, minimum and maximum bounds within a task completion.

On the other hand, a distributed computing framework is tremendously required for large processing loads in applications with multi-tasks. Hence, the prediction and design of completion times are critical to observing real-time processing with time-varying task communication.

4.3   Real-time embedded systems

The development of embedded systems with integrated micro-controllers has been accelerated in order to replace conventional electronic control systems. Embedded real-time systems are an important prerequisite for a wide range of applications such as those in smart industry, manufacturing automation, smart city and wireless communication. They act as an enabling technology, allowing systems to sense their environments and directly influence them through controlled and timely actions. A real-time embedded system plays the role of an intelligent part in a well-designed large system, which consists of mechanical component, embedded computational controller and human–machine interface. In fact, the variety of real-time applications of embedded systems and the size of real-time system markets have significantly expanded in the last few years and are expected to continue to grow in the era of 5G Internet-of-Thing (IoT) and industry 4.0. Digital information technology will see more dependence on 'embedded systems', not only in safety-critical scenarios such as disaster management, automotive devices, aerospace and medical devices but also in wireless communications or smart environment, thus creating a significant boost to the economy.

Intelligent decisions and predictable services are required of real-time application tasks in a real-time operating system. In this section, we focus on task management, inter-process communication, time management and error detection [71] in real-time embedded systems. Task management refers to the provision of a dynamic environment for the initialisation, execution and termination of application tasks within a host. Inter-process communication is to ensure information exchange among concurrently executed tasks to achieve progress. Flexible time management services are provided by the operating system to simplify or reduce the computational operation of the application software in real-time applications. Error detection must be introduced in a well-designed, real-time system for monitoring task execution time and unexpected interruptions, and detecting fail-silent nodes.

Below are the aspects of real-time embedded operating systems [73] that we need to consider:

- For networked embedded systems, middleware and platforms are deployed to deal with the complexity of computing, communication and control, while simultaneously providing effective distribution of low-cost resources. As a result, the design, programming, optimisation processing and system maintenance will become easier. Their scalability and self-organising platforms provide efficient services for ad-hoc networking and mastering complexity through novel techniques, and advanced computing and control.
- Methods and tools for system design are carefully planned and implemented in the development of software components with an emphasis on the correct handling of complex real-time constraints. This approach can combine computational models and composition methods, holistic design addressing event and time constraints, interface technologies in hardware and software addressing real-world and legacy issues, and techniques and integrated validation tools to ensure ultra-stable, dependable embedded systems.
- Hybrid system theories are emphasised in real-time systems for advanced controls including both constraints and switching modes in nonlinear processes. Advanced controls for networked embedded systems focus on networked autonomy and fault-adaptive control and management, as well as on reasoning, behaviour, global performance and robustness.

The combination of critical requirements, limited resources, concurrent environmental entities, high scalability requirements and high programming levels together with distributed processing presents specific problems for system engineering. Embedded real-time systems are now recognised as a distinct discipline, which has its own body of knowledge and theoretical foundations. The next generation of embedded real-time systems may expose many greater demands than those that are currently placed on them. Obviously, there is a need for coordinated research, technology transfers and educational programs with a view to providing support for the engineering of intelligent, flexible and dependable embedded real-time systems.

4.4 Real-time embedded convex optimisation

In this section, we focus on convex optimisation in real-time embedded systems. Real-time embedded optimisation is used in many areas such as signal processing, automatic control, real-time estimation, real-time resource allocation and fast decision-making in trading. The considerations of *"real time"* and *"embedded"* mean that optimisation algorithms execute much faster than a typical or generic method without any human intervention or action in the loop [74]. The embedded system using real-time optimisation algorithms is rapidly updated and fully automated with newly arriving data or changing conditions in time intervals that are measured in microseconds, milliseconds and few seconds for small-, medium- or large-size problems, respectively. This approach will be beneficial in solving optimisation problems with constantly updated data under a hard real-time deadline.

Following the continuous advances in computer science that offer more powerful processing, high-level programming languages and robust data analysis, solving times in the implementation of convex optimisation algorithms have been dramatically reduced. This entails new possibilities. Many optimisation problems which 20 years ago required hours to solve, can now take only microseconds. More exciting is that real-time convex optimisation is embedded directly in signal processing algorithms on real system running online and with strict real-time deadlines, even at low rates of tens of kilohertz of computer's performance [74]. By the flexible design of real-time algorithms, many optimisation problems are now potentially solved in one step or every few steps under time-bound constraints. It is imagined that more real-time signal processing algorithms will involve embedded optimisation in the future for a large number of real-time applications. In [74], the authors described two recent advances for the design and implementation of real-time convex optimisation algorithms. The first approach is to simplify problem specifications by allowing the transformation of problems in standard form to an automated platform using disciplined convex programming, thus will enable rapid application prototyping. The second advance is to automatically generate convex optimisation code by a customised solver from a high-level description of the problem family at a required high speed.

4.4.1   Disciplined convex programming

Disciplined convex programming (DCP) is an effective methodology to organise and implement parser-solvers for a natural form of convex optimisation problems. The natural form will be declared by setting up optimisation variables, defining objectives and specifying constraints. All proposed functions (objective and constraint functions) have to follow certain rules and be built-in with appropriate DCP forms (i.e., convexity forms) before proceeding to the next step. When these requirements are conformed, the parser can easily verify the convexity of the problem and automatically transform it to a standard form for solving with a customised solver. The advantage of a parser-solver (i.e., CVX, CVXPY) is that it would be much clearer for a larger, more complicated problem. Integrating further functions to the problem, including additional convex terms to the objective and constraints, is simple.

A standard DCP form in convex optimisation, linear programming, is to present the problem as a linear model and solve it by a generic solver. However, modern convex optimisation applications are represented as nonlinear models. The crucial difference from the previous category is that the DCP form and processing speed are now of critical importance. Consequently, DCP forms need to be continuously updated for supporting many types of convex optimisation problems, meanwhile, the total processing time must also be reduced. The described parser-solvers are ideal for this purpose, i.e., they can reduce development time by freeing users from the transformation of their problem into the standard form required by the solver.

4.4.2   Code generation

Automatic code generation is designed to formulate and test a convex optimisation algorithm within a familiar high-level environment and an appropriate customised solver. With an automatic code generation system, a user, who is not an expert in solving convex optimisation problems, can analyse and process the problem, using various methods. Then an efficient code platform including auxiliary code and files is produced for the particular problem. This code can be embedded in the user's signal processing algorithm. Choosing a suitable problem format is important for the design and prototyping of a convex optimisation-based algorithm with good application performance. Then, testing and generating code become easier to implement under a significant reduction in the speed of the solver.

Chapter 5
Introduction to practical optimisation problems

Optimisation problems are essential to engineering. Each optimisation application is the balanced trade-off of logistical and design strategies. Currently, there is a huge gap between optimisation theory and optimisation implementation in practice, resulting from the lack of research into this area.

As shown in Figure 5.1, the decision of optimisation application in practice is the balanced design of model environment, problem characteristics (variables, objectives), appropriate methods and trade-off (amongst performance, accuracy and timing) in real-time applications.

5.1 Stochastic optimisation

Stochastic optimisation or stochastic programming has been considered as having the ability to offer industry-standard approaches for optimisation applications [75]. Unlike deterministic optimisation methods, stochastic optimisation methods deal with highly nonlinear, highly dimensional, uncertain systems with relevant system noise. Hence, stochastic optimisation plays a crucial role in practical optimisation problems with random objective function or constraints.

A general convex stochastic program takes the form of [76]

$$\underset{x \in \chi}{\text{minimise}} \quad \mathbb{E}\{f(x, \omega)\} \tag{5.1a}$$

$$\text{s.t.} \qquad \mathbb{E}\{g_i(x, \omega)\} \leq 0, \ i = 1, ..., I \tag{5.1b}$$

$$h_j(x) = 0, \ j = 1, ..., J \tag{5.1c}$$

where $x \in \mathbb{R}^n$ and $\omega \in \mathbb{R}^q$ are the stable variables and random variables, $f : \mathbb{R}^n \times \mathbb{R}^q \to \mathbb{R}$ is the convex stochastic objective function, $g_i : \mathbb{R}^n \times \mathbb{R}^q \to \mathbb{R}$ is stochastic inequality constraints and $h_j : \mathbb{R}^n \to \mathbb{R}$ is deterministic or stochastic equality constraints.

In stochastic optimisation, we consider random variables and the expectations of objective function and constraints. The expectation can be approximated by using simple Monte Carlo evaluation as follows:

$$\mathbb{R}\{f(x, \omega)\} \approx \frac{1}{K} \sum_{k=1}^{K} f(x, \omega_k) \tag{5.2}$$

where K denotes the number of ω samples.

Figure 5.1   How to choose a good (best) decision for real-time optimisation application

5.1.1   Analysis of stochastic optimisation

The fast-changing characteristics of signal processing and wireless systems, such as time-varying fading channels, instantaneous resource allocations, or communication rates between system layers and between actions [77], prompt the need for stochastic optimisation. Unlike classic deterministic optimisation with perfect information assumption, stochastic optimisation considers random variables and statistic information, and thus, perfect information of the system is not available.

Minimisation of loss functions (also known as performance measure, objective function, measure-of-effectiveness, fitness function or criterion) is crucial in stochastic optimisation. For instance, the inherent noise in the loss function represents the measurements of physical systems and simulations to approximate a stable criterion.

In stochastic programming, randomness in the search direction method is a major principle in developing algorithms (e.g., random search algorithms) and is also a useful property of stochastic optimisation. In stochastic optimisation, randomness is used in probability and statistics calculation as a random process to assign a numerical value to each possible outcome of an event space. These structures are very useful in probability and statistic theory, and effective implementation in stochastic optimisation for various applications. The appropriate randomness model of the problems may be beneficial in speeding up the convergence of stochastic algorithms or dealing with modelling errors.

5.1.2   Characteristics of stochastic optimisation

First, since inherent noise information always exists under the loss function, the statistical error in algorithm input and result output needs to be reduced by calculating the costs of function evaluation. Second, in high-dimensional problems, the exploitation of problem structures can deal with the limit of multiple-variable optimisation that often dramatically increases with the problem dimensions [75].

Third, the convergence of optimisation algorithms is sometimes very difficult to determine in stochastic optimisation. Without prior knowledge of the information about the problems, the solution yielded by the algorithm can belong to some uncertain region of solution, which may prompt the algorithm to stop when it is close to the optimal solution. The algorithm may be suitable for the problem at a specific period of time, but within a rapidly changing environment, this does not ensure that the algorithm is still appropriate for the problem during the next. Simply put, given the updated process we will need to restart the algorithm and try to find a new solution, then check the convergence again. Stochastic methods are to automatically update and estimate the control process such that it will be appropriate for the problems in the current environment.

Although flexible stochastic optimisation algorithms can bring some benefits to real-time applications, they should be considered carefully for adapting and updating the algorithm processes. A small calibration in the coefficient settings of the algorithm might lead to a huge difference in the desired performance and achieved solution. Machine learning or deep learning, which will learn the operating scheme of the stochastic problems and then choose the suitable settings for updating the algorithms, is an efficient method for stochastic algorithms with flexible control process.

Practical optimisation problems may involve realistic constraints, e.g., varying time, opportunistic scheduling and strict solving time deadline. These constraints are often encountered in deterministic and stochastic approaches. Traditional optimisation methods are efficient for deterministic objectives and constraints with perfect information assumed. Stochastic approach deals with stochastic objectives and constraints, which frequently appear and change suddenly. These stochastic objectives and constraints are often ignored in solving optimisation problems – the main reason is that there are only a few practical methods for stochastic optimisation.

There is always a trade-off amongst the efficiency, reliability and stability of an algorithm. In fact, one optimisation algorithm can be suitable for one or several problems while others are not and vice versa. As a matter of fact, there will never be a universal search algorithm for all optimisation problems. This indicates that both deterministic and stochastic optimisations are continuing to be utilised in solving real-life problems.

5.1.3   Popular stochastic algorithms

Deterministic algorithms have a significant advantage as being capable of finding a global optimum in small or medium-scale optimisation problems. Since they focus on finding the global solution and proving the convergence of their algorithms, they might not be useful for large-scale problems. To deal with highly complex and large-scale problems, stochastic optimisation algorithms, which generate and use random variables and random iterates, can be applied.

In real problems, computational complexity is very important to the design of optimisation algorithms. Exact methods in deterministic class are impractical for problems of large size or too complicated evaluation [78]. In fact, since many realistic optimisation problems are non-deterministic polynomial-time hardness (NP-hardness), it is often difficult to guarantee the optimal solution in polynomial time. Instead, heuristics and metaheuristics are approximate methods to approach the appropriate solution by using the iterative process of trial-and-error. In this section, we introduce some popular stochastic algorithms for stochastic optimisation problems used in [78] and the references therein.

The simplest method is random search, where random potential solutions (e.g., local solutions) are picked up and evaluated [79]. The one with the best performance among the selected potential solutions will be treated as the final solution via random search. The main drawback is that this algorithm based on random choice may not guarantee the global optimum and may lead to poor solutions. Another approach is simulated annealing [80], a probabilistic technique for approximating the global solution in a large search space. Based on a Metropolis series with the process of reducing variation, this method can build a candidate solution and evaluate it, then select a neighbour candidate solution, check and update the better solution. Simulated annealing is efficient for large-scale problems, but might be slow since they cannot estimate and provide the exact number of potential candidate solutions. A similar approach to simulated annealing is the search method, which tries to overcome local optima (inexact solutions) by executing more search directions with the acceptance of non-improving solutions. The selection of a non-improving move is a probability used in a calculation cycle to the set of neighbours, not a given neighbour as in simulated annealing. To avoid meaningless calculation cycles, Tabu Search executes a list T of useful moves which may consist of t ($t \leq$ T) last moves by using aspiration levels.

As there is a lack of approaches for NP-hard problems, it is also difficult to know a good method for finding a good solution. With a given set of evaluated trial training, evolutionary algorithms (EAs) and genetic algorithms (GAs) [81, 82] can accumulate the training set and update a new set of parameters for getting the next process with much better performance than random search. However, in some large-scale problems, the set of trial training is not easy to create given a huge amount of training data. Based on artificial evolution, EAs use the selection and replacement operators for forcing this issue. Naturally, the selection of a strong candidate often leads to a strong convergence at least to the local optimum. Then, replacing the strong candidate in the next set of parameters in the algorithms will get better performance and fast convergence. In programming, GAs try to evolve this concept for individual processes. It is very efficient for multi-objective problems as genetic programming to run and execute for each individual process.

Another efficient method is particle swarm optimisation (PSO) [83, 84], which measures the particle quantity and moves the particles around in the search space over their position and velocity, and optimises the problem by iteratively improving a candidate solution. To find the best possible solution, the collaboration of particles is implemented, in which each particle can know the best solution's position ever found by swarm (gbest) and its best solution ever found (pbest). By observing the inertia, the particle goes towards both the directions of gbest and pbest with their modified speed. Hence, in the initial step, each particle assesses its current position and velocity and finds its best solution personally; after a while, it also finds the best solution by the particle swarm. In the main algorithm, the system initiates randomly all the particles with their evaluation and sets their gbest. The algorithm consecutively updates and copies the gbest for all particles and moves them with the velocity calculation. The algorithm will stop after checking that the criterion is met.

The next stochastic algorithm, known as ant colony optimisation (ACO), is a probabilistic technique for solving problems by finding good paths through graphs [85,86]. This approach is based on an ant colony's natural behaviour and describes the implementation of virtual ants in computing programming. The algorithm will compute based on the behaviour of ants within their self-organised society. Artificial ant communication will be designed for getting global information in their environment under pheromone forms over varying time for stochastic optimisation following virtual paths. Another similar stochastic algorithm, which is based on the intelligent foraging behaviour of honey bee swarms, is called artificial bee colony (ABC) [87]. In particular, ABC was inspired by honey bees with their smart foraging behaviour. These bees execute an essential mission: search for rich food sources, especially close to their hive. This model is based on the mechanism for self-organising and collective intelligence. For ABC, an artificial forager bee colony (agents) will search for rich food sources (good solutions). Next, agents randomly search and discover a set of initial solutions and then iteratively improve them by moving towards better solutions using neighbour search mechanisms while abandoning poor solutions.

Last but not least, a Markov-type algorithm called Markov Chain Monte Carlo (MCMC) [88] is used for sampling a probability distribution by constructing a Markov chain. Markov chain describes a sequence of possible events in which the probability of each event depends only on the state attained in the previous event. MCMC can obtain a sample of the desired distribution by observing the chain after a number of steps. This process will continue until the sample matches the actual desired distribution.

5.1.4 Stochastic optimisation in wireless communication systems

In [77], several ergodic stochastic optimisation (ESO) algorithms are presented to optimise resource allocation of wireless communication and networking. These algorithms are to solve stochastic optimisation problems with time-varying random state of channels. They deal with non-strictly concave Lagrangian and nonconvex resource allocation constraints but guarantee the convergence of ESO algorithms.

A framework for infinite-horizon average-cost problems in multi-dimensional communication systems is proposed in [89]. With low-computational complexity, a virtual continuous-time system is introduced to approximate the relevant value function of the original discrete-time system. The control delay problem is optimised for the K-pair interference networks by a distributed stochastic optimisation framework. In [90], a joint optimisation scheme of relay assignment and power allocation is proposed to maximise the total sum rate in relay networks for orthogonal multiuser networks. To solve the nonconvex sum-rate maximisation, this work exploits a stochastic method by using MCMC to find the global solution with low complexity.

In wireless communication systems, a beamforming-based spatial precoding method, which reduces the downlink training overheads and channel state information (CSI) feedback, is based on a stochastic optimisation algorithm [91]. This offers a better bit error rate performance and lower overheads of downlink training and CSI feedback compared to the conventional method with fixed beamforming coefficients. To find the optimal beamforming coefficients, the proposed beamforming design applies evolutionary PSO to eliminate inter-user interference.

In [92] and the references therein, Lyapunov drift optimisation concepts are considered for optimising network time averages in stochastic optimisation. Examples described in this work are multi-hop routing, maximum throughput with network coding, energy and delay constraints, and dynamic decision-making. These often occur with uncertainty, varying time and random events in mesh networks (heterogeneous networks, small cells), cognitive radio networks, wireless sensor systems and smart-grid. Hence, the analysis and control of stochastic network systems is necessary. Some interesting missions in [92] include minimising time average power with stability constraints, maximising throughput subject to time average power constraints or maximising throughput-utility under time average power constraints.

The stochastic network operates in discrete time with varying time slot $t \in \{0, 1, 2, ...\}$. In each slot t, a control action affects the arrivals and departures of the network queues with the attribute vectors $x(t) = (x_1(t), ..., x_M(t))$, $y(t) = (y_0(t), y_1(t), ..., y_L(t))$ and $e(t) = (e_1(t), ..., e_J(t))$ where M, L, J are non-negative integers. In this case, Lyapunov drift is used to solve the stochastic programs. $L(t)$ and $\triangle(t) = L(t + 1) - L(t)$ are the Lyapunov drift functions which define the backlog squares and actual queues on slot t and the difference of Lyapunov function between two adjacent time slots. For example, minimising the Lyapunov drift is equivalent to minimising $\triangle(t)$ to guarantee time average constraints.

5.2   Large-scale optimisation

Large-scale linear/nonlinear optimisation is concerned with the numerical solution of optimisation problems expressed as large problems, where the possible number of variables, objective functions and/or constraint functions are large, i.e., greater than 100 [93].

$$\underset{x \in \mathbb{R}^N}{\text{minimise}} \quad f_{\mathcal{K}}(x) \tag{5.3a}$$

$$\text{s.t} \qquad g_I(x) = 0 \tag{5.3b}$$

$$h_{\mathcal{J}}(x) \geq 0 \tag{5.3c}$$

where $f_{\mathcal{K}} : \mathbb{R}^N \to \mathbb{R}^K$, $g_{\mathcal{I}} : \mathbb{R}^N \to \mathbb{R}^I$ and $h_{\mathcal{J}} : \mathbb{R}^N \to \mathbb{R}^J$. Therein, $\mathcal{K} = \{1, ..., K\}$, $\mathcal{I} = \{1, ..., I\}$, $\mathcal{J} = \{1, ..., J\}$ and N, K, I, J are large.

The solution to such problems is likely to require large-scale linear algebra calculations at each iteration; moreover, the valuation of the objective function, constraints and their derivatives may be expensive due to large problem models.

5.2.1   Large-scale unconstrained optimisation

The general optimisation problem (5.3) without constraints (5.3b) and (5.3c) does not exist. It is a large-scale problem if the number of variables (N) and objective functions $(\mathcal{K})$ is also large. With an initial point $\mathbf{x}^{(\kappa)}$, the optimiser of the objective function can be found by a simple differentiable process without requiring trust region as in constrained problems.

If the objective functions $f_{\mathcal{K}}(\mathbf{x})$ are convex, line search method is useful to implement with

$$f_{\mathcal{K}}^{(\kappa+1)}(\mathbf{x}) = f_{\mathcal{K}}(\mathbf{x}^{(\kappa)} + \triangle^{(\kappa)}) \tag{5.4}$$

where $\triangle^{(\kappa)} = \alpha^{(\kappa)} d^{(\kappa)}$ is the optimiser step depending on the line search step-size $\alpha^{(\kappa)}$ and the direction $d^{(\kappa)}$ at the κth iteration. It is obviously that $f_{\mathcal{K}}^{(\kappa+1)}(\mathbf{x})$ is always a better feasible point than $f_{\mathcal{K}}^{(\parallel)}(\mathbf{x})$.

The next mission is to ensure the convergence of the objective function in (5.3a) by improving the optimiser step. Therefore, the computational cost of large-scale unconstrained convex optimisation problems is dominated by the computation values and required derivatives of objective functions at the steps of improving the optimiser.

Computation of Derivatives: With modest derivative requirements, the derivatives are computed very well. Furthermore, second-order (second derivative) can be commonly exploited by automatic differentiation and partial separability methods [93]. Tools of automatic differentiation can provide a significant efficiency computational possibility compared to previous methods in computing gradients and Hessian-vector products under a specific function. They are available as standalone software or within modelling and programming languages. In some cases, partial separability allows accurate structured gradient approximations (by finite-differences) and Hessian approximations (by secant formulas) to efficiently compute the partially separable structure or the specified structure of second derivatives.

Computation: Computation is also significantly complex if the problem consists of a large number of variables even when the derivatives of all functions are available. The required computations often implement with the quasi-Newton model

$$f(x) + d_k^T \nabla_x f(x) + \tfrac{1}{2} d_k^T \mathbf{H}_k d_k,$$

where $\mathbf{H}_k$ is symmetric positive definite by $\nabla_x f(x)$. Traditionally, the conjugate gradient method or Cholesky factorisation can be applied since $\mathbf{H}_k$ is positive definite. Cholesky factorisation is efficient for finite-difference gradients with high accurate evaluations since it only requires possible preconditioned residuals. On the other hand, when the Hessian $\mathbf{H}_k$ is indefinite, it is needed to modify factorisation or improve conjugate gradient process. For large-scale unconstrained optimisation, an alternative truncated-quasi-Newton is a breakthrough for dealing with high computational costs [93].

5.2.2   *Large-scale constrained optimisation*

The optimisation problem (5.3) with the existing (5.3b) and/or (5.3c) constraints is called a large-scale constrained optimisation. Depending on the type of objective function and constraint models, we can generally categorise large-scale constrained optimisation problems into large-scale linearly, nonlinearly and bound-constrained optimisation.

Large-scale linearly constrained optimisation: As its name suggests, this is a typical optimisation problem with large-scale linearly constraints. For example, a basic constrained problem is given as follows:

$$\underset{x \in \mathbb{R}^N}{\text{minimise}} \quad f(x) = c^T x + \frac{1}{2} x^T \mathbf{H} x \quad s.t. \ \mathbf{A}x = b \tag{5.5}$$

where $\mathbf{H}$ is symmetric positive definite and $\mathbf{A}$ is a matrix of full rank. This problem is called equality-constrained quadratic programming. An efficient factorisation method is the key solution for solving the above problem.

For more complex optimisation problems, large-scale nonlinearly and large-scale bound-constrained optimisation are very difficult practical problems that require a huge amount of computing resources for solving.

5.2.3   *Large-scale optimisation in the wake of big data*

Convexity in signal processing dates back to the dawn of the field, with problems like least-squares ones being ubiquitous across nearly all sub-areas. However, the importance of convex formulations and optimisation has increased dramatically during the last decade following the rise of new theories for structured sparsity and rank minimisation, and successful statistical learning models such as support-vector machines. These formulations are now employed in a wide variety of signal processing applications including compressive sensing, medical imaging, geophysics and bioinformatics [94].

The renewed popularity of convex optimisation places convex algorithms under tremendous pressure to accommodate increasingly large data sets and to solve problems of unprecedented dimensions. Internet, text and imaging problems (among a myriad of other examples) no longer produce data sizes from megabytes to gigabytes but rather from terabytes to exabytes. The practical utility of classical algorithms like interior point methods may not go beyond discussing theoretical tractability of the ensuing optimisation problems.

In response, convex optimisation is reinventing itself for Big Data where the data and parameter sizes of optimisation problems are too large to process locally, and where even basic linear algebra routines like Cholesky decompositions and matrix–matrix or matrix–vector multiplications that algorithms take for granted, are prohibitive. In contrast, convex algorithms also no longer need to seek high-accuracy solutions since big data models are necessarily simple or inexact.

5.2.4 Examples of large-scale optimisation

In [95], the authors considered a linear optimisation problem with individual linear programming (LPs) solved by using simplex method and matrix-free interior point method. The simulations were implemented on a computer with an Intel Core i7 3.07 GHz processor and 24 GB of RAM. The computational tests were performed using a Bell experiment under minimal critical visibility of GHz state as

$$\min v_c = 2^{\frac{1-n}{2}} \tag{5.6}$$

where n is the number of observers. Table 5.1 shows a comparison of the performance and solving time for problem (5.6) in two considered approaches. The n column represents the number of observers in a Bell experiment with 2 observables per observer. Minimal angles for the GHZ state were used in each case.

	Problem		Simplex method		Interior point method	
n	Name	Size	Time	Solution	Time	Solution
4	256	81	≈ 0 s	0.354	≈ 0 s	0.354
5	1 k	243	0.02 s	0.250	0.08 s	0.250
6	4 k	729	0.93 s	0.177	0.87 s	0.177
7	16 k	2 187	1 m 13 s	0.125	11.88 s	0.125
8	64 k	6 561	6 h 51 m	0.088	3 m 22 s	0.088
9	256 k	19 683	>>24 h	0.063	28 m 38 s	0.068
10	1 m	59 049	N/A	0.044	1 h 34 m	0.051

Although the interior point method is faster than the simplex method for larger problems, its solving time grows very quickly with the problem size.

5.3 Multi-objective optimisation

5.3.1 Definition of multi-objective optimisation

A general multi-objective optimisation problem is posed as follows:

$$\min/\max_{x} \quad F(x) = [F_1(x), ..., F_k(x)]^T \tag{5.7a}$$

$$\text{s.t.} \quad g_j(x) \leq 0, j = 1, ..., m \tag{5.7b}$$

$$h_l(x) = 0, l = 1, ..., n \tag{5.7c}$$

where k is the number of objective functions, m is the number of inequality constraints, n is the number of equality constraints and x is the vector of complicating design variables (also called decision variables). The feasible design space X (e.g., feasible decision space or constraint set) is defined as

$$\boxed{max\ F(x) = max\{f_1(x), -f_2(x)\}}$$

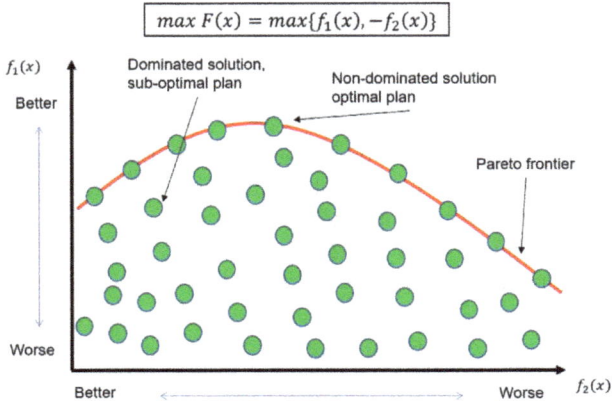

Figure 5.2   A Pareto optimal set for a multi-objective optimisation problem

$\{\mathbf{x}|g_j(\mathbf{x}) \leq 0, j = 1, ..., m; h_l(\mathbf{x}), l = 1, ..., n\}$. The feasible criterion space $\mathbf{Z}$ (e.g., feasible cost space) is defined as the set $\{F(\mathbf{x})|\mathbf{x} = \mathbf{X}\}$. Feasibility implies that no constraint is violated. Attainability implies that a point in the criterion space maps to a point in the design space. Consequently, even with an unconstrained problem, only certain points in the criterion space are attainable. In contrast to single-objective optimisation, for a multi-objective problem, there is no single global solution. And it is often necessary to determine a set of points that all fit a predetermined definition for an optimum. The predominant concept in defining an optimal point is that of Pareto optimality, which is defined as follows. A point, $\mathbf{x}^* \in \mathbf{X}$, is Pareto optimal if there does not exist another point, $\mathbf{x} \in \mathbf{X}$, such that $F(\mathbf{x}) \leq F(\mathbf{x}^*)$, and $F_i(\mathbf{x}) \leq F_i(\mathbf{x}^*)$ for at least one function under the considered minimisation problem. All Pareto optimal points lie on the boundary of the feasible criterion space $\mathbf{Z}$. Figure 5.2 illustrates a Pareto boundary for a simple multi-objective optimisation with two contradicting objectives $\{f_1(\mathbf{x}), f_2(\mathbf{x})\}$.

5.3.2   Example of multi-objective optimisation

We consider the energy efficiency (EE) maximisation problem of a multicell network where each of the base stations accompanied by a large-scale antenna array (e.g., up to several hundred) serve K single-antenna users. Based on the design of the transmit power $p = [p_{m,k}]_{m=1,k=1}^{M,K}$, let us define the objective functions as:

$$f_1(\mathbf{p}) = \sum_{m=1}^{m}\sum_{k=1}^{k} r_k(\mathbf{p}) \tag{5.8}$$

$$f_2(\mathbf{p}) = P_{tot}(\mathbf{P}) \tag{5.9}$$

where $r_k(\mathbf{p})$ represents the information throughput function at user k (e.g., semi-concave or nonconcave function due to interference) and $P_{tot}(\mathbf{p})$ represents the power consumption during transmission at the base stations (e.g., affine function or non-affine function due to the nonlinear power model). One of the interesting multi-objective optimisation problems in wireless networks is the network EE maximisation followed by the joint optimisation of two contradicting objectives $f_1(\mathbf{p})$ and $f_2(\mathbf{p})$. Accordingly, the EE optimisation problem is defined as

$$\max_{\mathbf{p}} \quad EE(\mathbf{p}) = \{f_1(p), -f_2(\mathbf{p})\} \text{ or } EE(\mathbf{p}) = \frac{f_1(\mathbf{p})}{f_2(\mathbf{p})} \tag{5.10a}$$

$$\text{s.t.} \quad r_k(\mathbf{p}) \leq \bar{r}_k, k = 1, ..., K \tag{5.10b}$$

$$P_{tot}(\mathbf{P}) \leq P_{tot}^{max} \tag{5.10c}$$

where (5.10b) and (5.10c) represent the throughput requirement (i.e., $\bar{r}_k$ is the throughput threshold at user k) and power budget constraints (i.e., P_{tot}^{max} is the maximum power consumption), respectively.

5.4 Integer programming and combinatorial optimisation

Combinatorial (or discrete) optimisation is the branch of optimisation in applied mathematics and computer science that is related to operations research, algorithm theory and computational complexity theory and sits at the intersection of several fields, including artificial intelligence, mathematics and software engineering [96].

All these problems, when formulated mathematically as the minimisation or maximisation of a certain function defined in some domain, have a commonality of discreteness. Combinatorial optimisation algorithms (exact algorithms and meta-heuristics) solve problems that are believed to be hard in general, by exploring the usually large solution search space of these instances. These algorithms achieve this by reducing the effective size of the space, and by exploring that space efficiently.

A popular model of combinatorial optimisation problems is the design of efficient integer programming and the availability of advanced computers. In the last decade, the use of integer programming models has increased dramatically. Problems with up to thousands of integer variables and more can be solved to optimality in the near future.

On the other hand, combinatorial algorithms often require a lot of resources to run. The two most important metrics for measuring the quality of a combinatorial algorithm are its execution time and the memory space that it uses. The first parameter is expressed in terms of the number of instructions necessary to run the algorithm. The use of the number of instructions as a unit of time is justified by the fact that the same program will use the same number of instructions on two different machines but the time taken will vary, depending on the respective speeds of the machines. The second parameter corresponds to the number of memory units used by the algorithm to solve a problem. The complexity in space is a function that associates an order of magnitude of the number of memory units used for the operations necessary for the solution of a given problem with the size of an instance of that problem. These advances have been made possible by developments in hardware, software and strategy design.

5.4.1 Branch-and-bound methods

The branch-and-bound (BnB) methods, also known as implicit enumeration approach in combinatorial optimisation, proceed using separation and sequential evaluation or divide and conquer method. Another name is the tree enumeration methods, implying separating and pruning without limiting pruning to the fruits of the evaluation. This can be a more accurate description of the characteristics of BnB methods. These methods perhaps belong to the exact approach since they provide one or all of the optimal solutions for various optimisation problems.

The common principle of BnB methods often involves an exhaustive enumeration with a fundamental difference. This enumeration can be made at the least exhaustive possible by the developer. Their principle has a common feature: constructing a root tree A (called the *search tree*); each vertex S of A represents a subset $\Omega(S)$ amongst the solution set Ω of the problem. The root R of A is associated with Ω itself: $\Omega(R) = \Omega$. The development of A from the root is done with the help of three main ingredients [96]: (i) a principle of separation, (ii) principles of pruning, generally the use of evaluation functions and bounds and (iii) a development strategy for the tree.

5.4.2 Dynamic programming

The title of dynamic programming is justified more by parallelly developing with that of linear programming. This approach is an optimisation method that, by proceeding the implicit enumeration of the solutions, allows us to solve sequential decision problems efficiently. More generally, it consists of a strategy with two essential parts for tackling optimisation problems including (i) breaking down the problem into a sequence of problems and (ii) establishing a recurrence link between the optimal solution of the problems.

5.5 Real-time optimisation problems

Each optimisation application is a balance and trade-off embedded in the analysis and design by developers. However, there is a huge gap between the optimisation theory and optimisation implementation in their applications. In addition, there is a lack of research to link the theory and implementation of optimisation. The decision of optimisation application in reality is a complex decision-making process which involves environment models, characteristics of the problem (variables, objectives), appropriate methods and trade-offs for an expected solution [97].

The role of processing time, which depends on how fast optimisation problems are solved, is very crucial in signal processing and wireless communication. Currently, traditional convex optimisation methods are still expensive and extremely time-consuming. This calls for the development of novel strategies, which could be particularly beneficial when applied at the times of mission-critical communications such as natural disasters. A real-time optimisation process has to respond to the following questions:

- How to deal with the execution time of optimisation algorithms in real-time scenarios?
- How to overcome large-scale problems and/or high-computational complexity of problems in online applications?
- How to generate and embed efficient algorithms for real-time optimisation framework in the real systems?

In the context of optimisation theory, a mathematical model optimisation problem from the real world can be transformed into a practical online optimisation problem. Then this problem can be programmed in embedded systems with the online solutions. The online solutions found should be appropriate to the problem and satisfy the accuracy of performance. Meanwhile, the strict time deadline should be adhered to in real-time implementation. As a result, there are many challenges in real-time optimisation in the sense that 'faster is better' [74]. For wireless communication, fast optimisation is used to calibrate some weights to improve the transmission strategy, optimise network performance or other objectives. However, practical optimisation problems in wireless networks are frequently nonlinear programming, large-scale problems with massive amounts of data (big data analytics), multi-objective and/or high-complex problems, stochastic problems, sparsity or low-rank problems (channel assignments) and dynamic programming (hybrid resource allocation problems).

5.6   Introduction to methodologies of real-time optimisation

Although embedded optimisation programming has proven promising in reducing solving time for optimisation problems, modern problems are continuously arising in both complexity and scale. That is, embedded programs alone cannot effectively tackle the issues in current and future optimisation contexts. In what follows, we present some potential approaches for real-time optimisation applications in signal processing and wireless networks and resilient Internet-of-Thing (IoT). Our aim is to introduce the key approaches, their collaboration and open questions related to the use of real-time optimisation framework for solving modern optimisation problems across several fields in real systems. The first approach is first-order methods. Having the benefit of gradient methods in optimisation concept, first-order methods use first derivatives to design optimisation algorithms for finding the optimum. This approach is simple, efficient and suitable for both convex and nonconvex optimisation problems, linear and nonlinear programming and large-scale and complex problems. Thus, it has been considered as the root of real-time optimisation algorithms [98–100]. We will present the classic first-order methods and their significant improvement by their variants, accelerated first-order, proximal approaches and

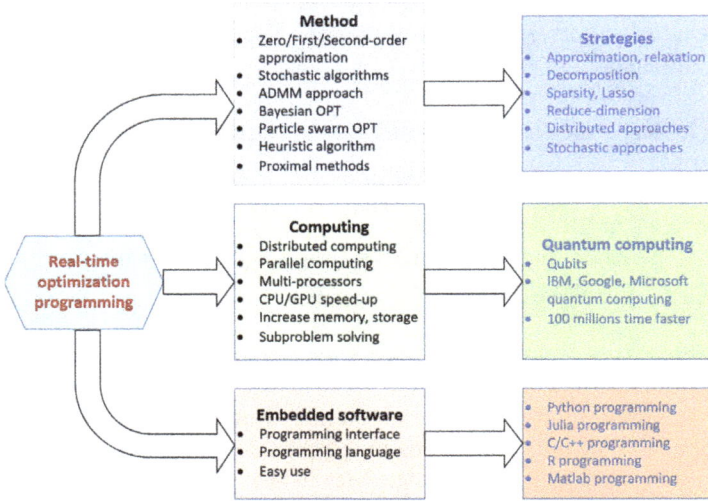

*Figure 5.3   Potential approaches for real-time embedded optimisation
            programming*

stochastic gradient methods as considered in [101]. The second approach is parallel
and distributed computing platforms for modern wireless communication problems,
e.g., large-scale problems, big-data transfers, real-time signal processing control and
hybrid resource allocation with multi-objective problems [102–105]. Another top
approach is the interplay of machine learning (deep learning models) and optimisation algorithms. This approach is exploited in modern wireless communication as an
emerging technique for real-time optimisation applications [106–110]. Faced with
many challenges of 5G networks and beyond, the collaboration of machine learning and optimisation techniques is now the hottest trend of practical optimisation in
real-time applications.

Chapter 6

First-order methods for real-time optimisation

Although gradient-based methods are used widely in real-time optimisation, they may not be efficient when the gradient and Hessian information is difficult to evaluate, e.g., there are no explicit function forms or non-differentiable functions. To overcome this issue, derivative-free methods such as evolutionary algorithms or swarm intelligence-based algorithms can be used for finding the optimum from potential solution candidates [111, 112]. Otherwise, higher-order approaches based on higher-derivative information [101, 113] such as second-order algorithms are appropriate when high-accuracy performance is required, i.e., the exact optimum of the problem is necessary.

As for higher-order algorithms, the number of derivative operations or the evaluation of costs is also largely depending on the problem size in large-scale, big-data analysis or multi-objective optimisation. Higher derivatives are often difficult to evaluate, even in second-order operation for complex objectives. On the other hand, while zero-order algorithms can achieve the exact optimal solution, they often converge very slowly without guaranteeing the exact optimal solution and convergence rate. In summary, with roots in derivative context and convexity optimisation programming, first-order methods are the most popular in the context of optimisation [1, 98, 101, 114, 115].

6.1 An overview of first-order methods

We begin by discussing first-order methods for solving optimisation problems. Here, 'first-order' refers to the first-order derivatives of the function that these techniques require [98, 100, 101, 114, 116]. There are many benefits of the first-order methods in the context of optimisation applications.

First, solving nonconvex optimisation problems in signal processing and wireless communication is a challenging task, especially now that the number of complicated problems and nonconvex optimisation programming has significantly increased. Adding real-time requirements further makes it more difficult to solve these optimisation problems. To overcome this issue, nonconvex optimisation problems will first need to be transformed into convex ones, which are then addressed by using various convex optimisation algorithms. This process can be simply implemented by using first-order techniques. Then, the local optimum can be easily found in convex optimisation problems and, thus, the globally optimal point is also found out by the concept of convexity. Together with the explosion of wireless communication, first-order convex optimisation will become the most potential tool for the design, analysis, deployment of wireless communication systems.

Second, although they are often numerical solutions of low or medium accuracy, first-order methods are efficient for many optimisation problems, even large-scale operation. This is because these methods exhibit nearly dimension-independent convergence rates. In fact, first-order methods are iterative algorithms that only exploit information of the objective function and its gradient (sub-gradient). They require minimal data information of the problem. Thus, they are very simple and low-cost iterative schemes that are suitable for large-scale problems when high accuracy is not crucial.

Moreover, they can also handle non-smooth problems by making use of proximal mapping principle as discussed later. Finally, these techniques provide flexible frameworks for distributing optimisation tasks and performing tasks in parallel and, thus, they are ideal for distributed and parallel computation, from synchronous parallel algorithms (centralised) to scalable asynchronous algorithms (decentralised).

Some popular first-order algorithms are fixed-point methods [117, 118], coordinate descent [98, 119], alternating minimisation [120], gradient and subgradient methods, stochastic gradients [101, 121], proximal algorithms and augmented Lagrangians and splitting [122]. They are used in many applications in real world such as clustering analysis, statistical estimation, machine learning and deep learning, signal and image processing and wireless communication systems. In the following, we introduce the basic of first-order methods and provide some feasible approaches for the improvement of first-order methods in real-time applications.

A general optimisation problem (P1) is given as

$$\underset{x}{\text{minimise}} \quad f(x)$$

$$\text{s.t.} \quad g_i(x) \leq 0, i = 1, ..., I$$

$$h_j(x) = 0, j = 1, ..., J,$$

Considering the above problem as the model of convex problem (1.5), a feasible point can be found by using the gradient method from a feasible $(k-1)$th iterative point

$$\mathbf{x}_k = \mathbf{x}_{k-1} - \alpha_k \nabla f(\mathbf{x}_{k-1}), \quad k = 1, 2, ... \tag{6.1}$$

where α_k is the stepsize. The value of α_k can be fixed or determined by a line search technique. The gradient method proposes the convergence rate of $\mathcal{O}(1/k)$.

On the other hand, when problem (P1) is a nonconvex problem, there are one or several functions which are nonconvex, including the objective and constraint functions. For instance, we assume that the objective function $f(x)$ is a non-convex form but differentiable and having a concave form. To this end, we can use a first-order method to handle the objective function $f(x)$. A general first-order approximation for the nonconvex objective function is expressed as

$$f(x) \leq \hat{f}(x) \tag{6.2}$$

where

$$\hat{f}(x) = f(y) + \nabla f(y)(x - y),$$

y is a given point of $f(x)$.

For vector and matrix presentation, one has

$$f(X) \leq \hat{f}(X) = f(Y) + \langle \nabla f(Y), (X - Y) \rangle \tag{6.3}$$

where $\langle X, Y \rangle$ denotes the inner product of (X, Y). The key strategy of these approximations is that the functions on the right-hand side of (6.2) and (6.3) will turn (P1) into a convex problem.

First-order embedded optimisation programming:

Here, we will discuss the embedded programming for real-time optimisation. This first-order approach is beneficial for complicated and large-scale nonconvex problems. We start with problem (P1), which is supposed as a nonconvex problem because of the nonconvexity of the objective function. By using a concavity of $f(x)$ and first-order approximation approaches (6.2) and (6.3), we can transform the objective function $f(x)$ to its upper bound with a convexity form. One has

$$f(x) \leq \hat{f}^{(\kappa)}(x) = f(x^{(\kappa)}) + \nabla f(x^{(\kappa)})(x - x^{(\kappa)}) \tag{6.4}$$

where $x^{(\kappa)}$ is the feasible point at the κth iteration.

At the κth iteration, the following convex program is solved to generate the next feasible point

$$\underset{x \in \chi}{\text{minimise}} \quad \hat{f}^{(\kappa)(x)} \tag{6.5a}$$

$$\text{s.t} \qquad g_i \leq 0, \ i = 1, ..., I, \tag{6.5b}$$

$$h_j(x) = 0, \ j = 1, ..., J, \tag{6.5c}$$

In the next iteration $\kappa + 1$, we define $x^{(\kappa+1)} = x^*$ where x^* is a temporary saturation point of problem (6.5). It is obvious that $x^{(\kappa+1)}$ is a better feasible point than $x^{(\kappa)}$ due to the convexity of the approximated function $f^{(\kappa)}(x, x^{(\kappa)})$:

$$f(x^{(\kappa+1)}) \leq f(x^{(\kappa)}). \tag{6.6}$$

Thus, in Algorithm 1 we propose a computational procedure using first-order approximation for solving the large-scale nonconvex optimisation problem (1.5), as follows:

Algorithm 1 Path-following algorithm for embedded optimisation programming

---

1: **Initialisation**: Set $\kappa := 0$. Set a feasible point $x^{(0)}$ for problem (P1).
2: **Repeat**
3:    Solve the problem (6.5) for its optimal point x^* using automatic code-generated in embedded programming.
4:    Set $\kappa := \kappa + 1$. Then, set $x^{(\kappa+1)} = x^*$.
5: **Until** A convergence of $f(x)$.

6.2 Accelerated first-order approaches

The main problem of first-order methods is that they often have low convergence rate when optimisation problems become very large-scale and require high accuracy. In the first instance, we provide a relaxation technique by using the second-order approximation for the function $f(x)$ as below

$$f(x) \approx f(y) + \nabla f(y)^T (x - y) + \frac{1}{2}(x - y)^T \nabla^2 f(y)(x - y) \tag{6.7}$$

where $\nabla^2 f(.)$ denotes the second derivative of function $f(.)$.

However, as discussed before, this solution depends on second-order evaluation, which is sometimes difficult in complex functions and large-scale problems. To overcome this issue, we use accelerated first-order methods, whose main goal is to improve the convergence rate of first-order methods.

By following the Lipschitz continuity, we have:

$$\|\nabla f(x) - \nabla f(y)\| \leq L(f)\|x - y\| \quad \forall \, x, y \tag{6.8}$$

where $L(f) > 0$ denotes the Lipschitz constant of ∇f.

For any, $L > L(f)$, some properties of Lipschitz continuity are as follows:

$$\|\nabla f(x) - \nabla f(y)\| \leq L\|x - y\| \quad \forall \, x, y \tag{6.9a}$$

$$\frac{1}{L}\|x - y\| \leq \|\nabla f(x) - \nabla f(y)\| \leq L\|x - y\| \quad \forall \, x, y \tag{6.9b}$$

One has

$$\mu \mathbf{I} \preceq \nabla^2 f(x) \preceq L\mathbf{I} \tag{6.10}$$

where $\mathbf{I}$ denotes the identity matrix and appropriate μ and L.

Using the inequality $\nabla^2 f(x) \preceq L\mathbf{I}$, then one has

$$f(y) \leq f(x) + \nabla f(x)^T (y - x) + \frac{L}{2}\|y - x\|^2$$

Using that $\nabla^2 f(x) \succeq \mu \mathbf{I}$, then one has

$$f(y) \geq f(x) + \nabla f(x)^T (y - x) + \frac{\mu}{2}\|y - x\|^2$$

Following the simple approach as shown in [123], by introducing a new control parameter β, the accelerated gradient descent can easily improve the convergence rate of classic gradient descent:

$$x_{k+1} = x_k - \beta \nabla f(x_k) \tag{6.11}$$

which obtains the convergence rate of $\mathcal{O}(1/k)$. By adding one more step, the accelerated gradient descent is as follows:

$$y_{k+1} = x_k - \beta \nabla f(x_k) \tag{6.12a}$$
$$x_{k+1} = (1 - \lambda_k)y_{k+1} + \lambda_k y_k \tag{6.12b}$$

with appropriate choice of λ_k for each iterate. The accelerated gradient descent can obtain the convergence rate of $\mathcal{O}(1/k^2)$.

Iterative shrinkage-thresholding algorithms (ISTA) and fast iterative shrinkage-thresholding algorithms (FISTA) are two early methods representing the accelerated approach for first-order methods [124]. Now we focus on the following general large-scale non-smooth convex problem

$$\min_x F(x) = f(x) + g(x) \tag{6.13}$$

where f, g are convex functions, with g frequently non-smooth. Based on the class of first-order methods, [124] confirmed that the ISTA behaves as

$$F(x_k) - F(x^*) \simeq \mathcal{O}(1/k) \tag{6.14}$$

where x^* is supposed as the optimal solution of $F(x)$. The ISTA provides the sublinear convergence rate like classic gradient descent.

Then, adopting the following first-order approximation and Lipschitz property, for $L > 0$, one has

$$\hat{F}_L(x, y) = f(y) + \langle x - y, \nabla f(y) \rangle + \frac{L}{2}\|x - y\|^2 + g(x) \tag{6.15}$$

Next, one has

$$p_L(y) = \operatorname{argmin}\{\hat{F}(x, y)\} = \operatorname{argmin}\{g(x) + \frac{L}{2}\|x - (y - \frac{1}{L}\nabla f(y))\|^2\} \tag{6.16}$$

Finally, for each iterate with $y := x_{k-1}$, we have

$$x_k = p_L(x_{k-1}) \tag{6.17}$$

Algorithm 2 sums up the ISTA with constant stepsize procedure.

Algorithm 2 ISTA with constant stepsize

1: **Initialisation:** L is a Lipschitz constant of ∇f. Choose feasible x_0 and $k = 0$.
2: **Repeat**
3: Set $k := k + 1$.
4: Using (6.17) compute x_k.
5: **Until** converge to a criterion, e.g., $\|x_k - x_{k-1}\|^2 \leq \epsilon$.

However, in some cases such as large-scale problems, the Lipschitz constant L is not always found since it depends on the second derivative operation. To deal with this issue, ISTA with a backtracking stepsize rule is proposed as in Algorithm 3 [124].

Algorithm 3 ISTA with backtracking stepsize

1: **Initialisation:** Set $L_0 > 0$ and $\eta > 1$. Choose feasible x_0 and $k = 0$.
2: **Repeat**
3: Set $k := k + 1$.
4: Find the smallest integer $i_k \geq 0$ such that $L_k = \eta^{i_k} L_{k-1}$ with the fact that

$$F(p_L(x_{k-1})) \leq \hat{F}_L(x_{k-1}, p_L(x_{k-1}))$$

5: Using (6.17) with L_k to compute x_k as $x_k = p_{L_k}(x_{k-1})$.
6: **Until** convergence to a criterion, e.g., $||x_k - x_{k-1}||^2 \leq \epsilon$.

Therefore, for a non-smooth problem as (6.13), the Lipschitz continuity and ISTA approximate the original problem to a relaxed form as (6.11) for applying gradient descent method with a convergence rate of $\mathcal{O}(1/k)$. To improve the convergence rate, FISTA is proposed so as to obtain the convergence rate of $\mathcal{O}(1/k^2)$. Two modified versions of Algorithms 2 and 3 with FISTA scheme are presented as Algorithms 4 and 5.

Algorithm 4 FISTA with constant stepsize

1: **Initialisation:** Set L (a Lipschitz constant). Choose feasible $y_1 = x_0$, $t_1 = 1$ and $k = 0$.
2: **Repeat**
3: Set $k := k + 1$.
4: Compute $x_k = p_L(y_1)$, $t_{k+1} = \frac{1+\sqrt{1+4t_k^2}}{2}$ and $y_{k+1} = x_k + (\frac{t_k-1}{t_{k+1}})(x_k - x_{k-1})$.
5: **Until** convergence.

Algorithm 5 FISTA with backtracking

1: **Initialisation:** Set $L_0 > 0$ and $\eta > 1$. Choose feasible $y_1 = x_0$, $t_1 = 1$ and $k = 0$.

2: **Repeat**

3: Set $k := k + 1$.

4: Find the smallest integer $i_k \geq 0$ such that $L_k = \eta^{i_k} L_{k-1}$ with the fact that

$$F(p_L(y_k)) \leq \hat{F}_L(y_k, p_L(y_k))$$

5: With L_k, compute $x_k = p_{L_k}(y_k)$, $t_{k+1} = \frac{1+\sqrt{1+4t_k^2}}{2}$ and $y_{k+1} = x_k + (\frac{t_k - 1}{t_{k+1}})(x_k - x_{k-1})$.

6: **Until** convergence.

For example, we consider the following problem [125]

$$\min_{x \in \mathbb{R}^n} f(x) := \log \sum_{i=1}^{m} \exp(a_i^T x + b_i) \tag{6.18}$$

with randomly generated data $m = 2000$, $n = 1000$ and the FISTA with constant stepsize is used. Some figures show the superior convergence of FISTA compared to the classic gradient descent for problem (6.18), as provided in [125].

6.3   Proximal methods for non-smooth problems

FISTA method can provide fast convergence rate and is simple to implement. However, for problems that contain non-smooth (non-differentiable) functions in objective functions and/or constraints such as regulariser terms, sparsity functions, absolute functions, norm functions or other complex functions, its derivative is difficult to obtain.

To address this, there are several efficient techniques including smooth approximation, reforming the non-smooth problem as a constrained problem, applying projections onto simple sets and proximal mapping or proximal-gradient (projected-gradient) method [122]. In this section, we will discuss several methods through illustrated examples.

First, smooth approximation [126] is suitable for simple non-smooth functions such as

$$|x| \approx \sqrt{x^2 + \epsilon},$$

where ϵ has a very small value.

$$\max\{a, b\} \approx \log(\exp(a) + \exp(b))$$

Table 6.1   Projections of some non-smooth functions

Non-smooth functions	Projections
$\max\{x,0\}$	$\operatorname{argmin}_{y\geq 0} \|y-x\|$
$\max\{l,\min\{x,u\}\}$	$\operatorname{argmin}_{l\leq y\leq u} \|y-x\|$
$x+(b-a^T x)a/\|a\|^2$	$\operatorname{argmin}_{a^T y=b} \|y-x\|$
$\tau x/\|x\|$	$\operatorname{argmin}_{\|x\|\leq\tau} \|y-x\|$
x if $a^T x\geq b$ or	$\operatorname{argmin}_{a^T y\geq b} \|y-x\|$
$x+(b-a^T x)a/\|a\|^2$ if $a^T x<b$	

$$\max\{0,x\} \approx \begin{cases} 0 & x\geq 1 \\ 1-x^2 & t<1 \\ (1-t)^2+2(t-1)(t-x) & x\leq t \end{cases}$$

Similarly, another smooth approximation known as 'projections' [127] can be used on simple feasible sets. Table 6.1 shows some simple projections of non-smooth functions onto simple sets.

The third technique for non-smooth optimisation problems is to reform non-smooth functions in objective functions as a constrained problem. Considering the convex problem as

$$\min_x f(x)+\lambda\|x\|_1 \tag{6.19}$$

where $f(x)$ is a convex function and the regulariser is also a convex but non-smooth function with sparsity $\|x\|_1\leq\gamma$. Introducing a new problem with $\|x\|_1\leq\gamma$, then the problem (6.19) is equivalent to

$$\min_x \quad f(x)+\lambda\gamma \tag{6.20a}$$
$$\text{s.t.} \quad \|x\|_1\leq\gamma \tag{6.20b}$$

An efficient method for non-smooth problems is to apply proximal mapping [125, 127]. Any non-smooth convex function is always presented by a proximal operator. For a convex function $h(x)$, the general proximal operator of h is as follows:

$$\operatorname{prox}_{th}(x)=\operatorname*{argmin}_{u\in E}\left(h(u)+\frac{1}{2t}\|u-x\|^2\right) \tag{6.21}$$

where $x,u\in\mathbb{R}^n$ and $t>0$.

For instance, given $h(x)=\|x\|_1$, then the proximal of h, prox_h is the soft thresholding (shrinkage) function as

$$\operatorname{prox}_{th}h(x)_i = S_t(x_i) = (|x|-t)_+\operatorname{sign}(x_i) = \begin{cases} x_i-t & x_i\geq t \\ 0 & |x_i|\leq t \\ x_i+t & x_i\leq -t \end{cases}$$

Table 6.2 shows some simple proximal mappings of non-smooth functions.

Table 6.2   Proximal mappings of some functions

Functions	Proximal mapping
$h \equiv c$ (constant function)	$\mathrm{prox}_{th}(x) = x$
$h(x) = <a, x> + b$ (affine function)	$\mathrm{prox}_{th}(x) = x - ta$
$h(x) = 1/2 x^T A x + b^T x + c,\ A \in \mathbb{S}^n +$	$\mathrm{prox}_{th}(x) = (I + A)^{-1}(x - tb)$
(quadratic function)	
$h(x) = \|x\|2$ (Euclidean norm)	$\mathrm{prox}_{th}(x) = (1 - t/\|x\|2)x\ if\ \|x\|2 \ge t$ or 0
	otherwise
$h(x) = \min_{y \in c}\|x - y\|2,\ C$ closed convex	$\mathrm{prox}_{th}(x) = \theta P_C(x) + (1-\theta)x,\ \theta = t/d(x)$ if
(Euclidean distance)	$d(x) \ge t$ or 1 otherwise

Considering the composite optimisation problem

$$\min_{x} g(x) + h(x) \tag{6.22}$$

where g is a smooth function (not necessarily convex) and h is a non-smooth (necessarily convex). Applying the first-order approximation with Lipschitz continuity, one has

$$x_{k+1} = \operatorname*{argmin}_{x} g(x_k) + \langle \nabla g(x_k), x - x_k \rangle + \frac{L}{2}\|x - x_k\|^2 + h(x) \tag{6.23}$$

Then, the proximal-gradient algorithm [125] with the proximal quadratic optimisation is shown as

$$x_{k+1} = \operatorname*{argmin}_{x} \frac{1}{2}\|x - (x_k - \beta\nabla g(x_k))\|^2 + \beta h(x) \tag{6.24}$$

$$x_{k+1} = \mathrm{prox}_{\beta}[x_k - \beta\nabla g(x_k)] \tag{6.25}$$

where $\mathrm{prox}_t(x) = S_t(x) = (|x| - t)_+\mathrm{sign}(x)$.

For example, $\min \frac{1}{2}\|y - Ax\|^2 + t\|x\|_1$ with the proximal of regulariser term as $\mathrm{prox}_t(x) = S_t(x) = (|x| - t)_+\mathrm{sign}(x)$, where $S_t(x)$ is the soft-thresholding function. Combining an accelerated gradient descent (e.g., FISTA) and the proximal algorithm above [128, 129], let $\mathcal{O}(1/k^2)$ be a feasible initial point, then

$$y_{k+1} = \mathrm{prox}_{\beta}[x_k - \beta\nabla g(x_k)]$$
$$x_{k+1} = (1 - \lambda_k)y_{k+1} + \lambda_k y_k$$

where $\lambda_k = \frac{1-t_k}{t_{k+1}}$, $t_k = \frac{1+\sqrt{1+4t_{k-1}^2}}{2}$, $t_0 = 0$. The purpose of combining accelerated gradient descent and proximal algorithm is two-fold. First, the non-smooth composite problem then can be handled by a proximal algorithm. Second, the convergence rate is accelerated to $\mathcal{O}(1/k^2)$ by FISTA method. The proof of convexity and more results were shown in [129].

6.4 Stochastic gradient methods

Optimisation problems in wireless networks are often large-scale, with a huge amount of data to be analysed and optimised at the same time. An efficient method that can be used is stochastic gradient descent (SGD) and its variant [101, 121, 130, 131]. SGD is an incremental gradient descent using stochastic approximation and iterative method for optimising the problem with large-scale operations.

Considering a big N problem as

$$\min_{x} \frac{1}{N} \sum_{i=1}^{N} f_i(x) \tag{6.26}$$

For classic (deterministic) gradient method, at the $k + 1$ iteration,

$$x_{k+1} = x_k - \alpha_k \nabla f(x_k) = x_k - \frac{\alpha_k}{N} \sum_{i=1}^{N} \nabla f_i(x_k)$$

Note that the computational iteration cost is linear in N since we need to evaluate N-derivative sequentially. The critical issue is that the computational cost might be prohibitive when N is very large. To deal with this, SGD will be used [122]. In the beginning, the process randomly selects a small set of f from $\{1, ..., N\}$. Then,

$$x_{k+1} = x_k - \alpha_k \nabla f(x_k) \approx x_k - \alpha_k f_i'(x_k)$$

where f_i' is the stochastic gradient of f based on the set of i. The estimate of the true gradient of f_i' is as follows:

$$\mathbb{E}[f_i'(x)] = \frac{1}{N} \sum_{i=1}^{N} \nabla f_i(x)$$

The full gradient approach by using N subgradients at the Nth iteration is

$$\nabla f(x_k) = \frac{1}{N} \sum_{i=1}^{N} \nabla f_i(x_k)$$

The SGD approximations with finite number of samples $\mathcal{B}_k$ are

$$\frac{1}{|\mathcal{B}_k|} \sum_{i \in \mathcal{B}_k} f_i'(x_k) \approx \frac{1}{N} \sum_{i=1}^{N} \nabla f_i(x_k).$$

Then,

$$x_{k+1} = x_k - \frac{\alpha_k}{|\mathcal{B}^k|} \sum_{i \in \mathcal{B}_k} f_i'(x_k).$$

Note that the iteration cost of SGD is independent of N. Thus, stochastic iterations are N times faster than classic GD. More importantly, this approach can be exploited in stochastic optimisation problems, which represent the optimisation in real wireless communication systems, and is introduced in Section 5.5.

However, SGD often generates much larger iterations since they do not take the full gradient of f, and, thus, it is hard to control the convergence of SGD. In summary, the stochastic gradient approach has a low iteration cost but slow convergence rate. Another

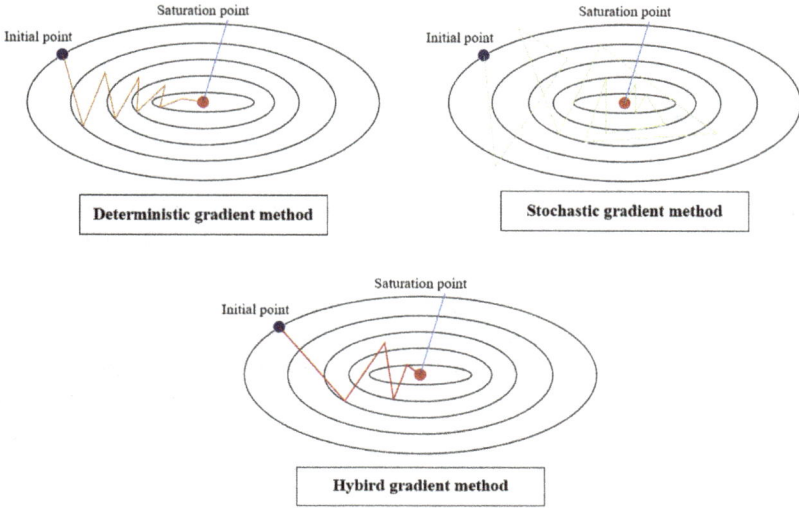

Figure 6.1   The convergence of deterministic and stochastic gradient method

point is that the stochastic gradient approach is superior in terms of solving time but is of very low accuracy (Figure 6.1).

SGD is an efficient optimisation method for unconstrained optimisation problems. For general optimisation problems with equality and inequality constraints, the hybrid methods that combine first-order GD and SGD methods are promising for real-time applications in wireless communications, as shown in Figure 6.2.

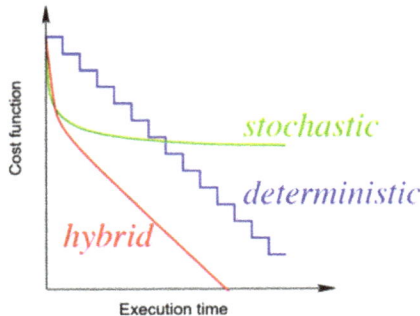

Figure 6.2   The hybrid approach of deterministic and stochastic gradient descent

6.5   Applications of first-order optimisation in 5G IoT

In wireless communication systems, first-order or first approximation optimisation methods have been widely used in resource allocation [132, 133], scheduling and routing schemes [134–136], interference alignment [137–141], physical layer security [142, 143], energy harvesting [144–146] and so on.

In our works [25, 51, 147–149], we applied first-order approximation methods for resource management of emerging scenarios in 5G wireless networks, e.g., small-cell, massive MIMO, massive MIMO heterogeneous networks (HetNets) and cell-free networks. Therein, the EE maximisation problem is often a nonconvex problem due to the fractional function (nonconvex class) of EE objective function and several nonconvex constraints. By using the first approximation, our proposed EE maximisation procedures have low complexity and are efficient to approximate the nonconvex EE problem. In [25, 150], the three-objective optimisation of EE, quality-of-service (QoS) requirement and user traffic loading were analysed and optimised for small-cell networks.

A joint linear precoder design and small-cell switching-off approach were proposed for improving network EE performance. Using a novel group sparsity, the small-cell BSs (SBSs) that have the ability to serve their users best are activated. Meanwhile, other SBSs that have a negligible contribution will be turned-off to further improve the system EE. In [51, 147, 148], low-complexity beamforming design procedures for resource allocation are provided for massive MIMO systems and their application in massive MIMO cell-free and massive MIMO HetNets. To deal with the large-scale size of beamforming design in massive MIMO systems, classic beamforming designs (conjugate, zero-forcing (ZF) and regulariser zero-forcing (RZF) beamformers) and power allocation techniques are collaborated for optimising the EE maximisation problem under practical constraints.

Novel optimisation approaches such as convex optimisation techniques and logarithm inequalities are proposed to deal with the classes of fractional programming in the presence of the EE objective. In the simulation results, these approaches always achieved a better performance compared to existing schemes in terms of EE. In the age of the IoT, hybrid resource allocations of massive MIMO systems are potential approaches for serving a larger number of users simultaneously [147, 149]. Simulation results showed that by combining a novel time schedule transmission and resource allocation, the number of users that are served can be much more than the number of antennas at the BS.

Although classic 'first-order' standards are popular in many areas of engineering, their main focus is on solving the problems. Thus there is a lack of works considering the computational complexity and execution time in the implementation of optimisation algorithms. This means that most previous works did not take account of practical problems in real-time applications. In realistic systems on real-time control, first-order methods might not be efficient since they often have a slow convergence rate and can only be adapted to solve problems at low accuracy. In the next chapters, to improve the solving performance of first-order methods, we will introduce novel approaches such as distributed/parallel computing approaches and an interplay of optimisation and machine learning.

Chapter 7

Distributed and parallel computing for real-time optimisation

In future wireless communication, various complex and interrelated problems may concur, albeit within a temporal sequence [104, 105]. Large-scale problems, big-data analysis, real-time predictive control, multi-objective problems and hybrid resource allocation of communication systems are some good examples. In this chapter, we have several reasons to discuss distributed approaches and parallel computing for optimisation of wireless networks, especially for real-time applications. Distributed and parallel computing is better suited for the modelling, simulation and understanding of complex, real-world problems. From the functional point of view, it is only a little difference between the implementation with a centralised architecture and that with a distributed architecture. However, a number of arguments are in favour of the distributed approach for the implementation of hard real-time systems. Two important aspects of distributed or parallel computing architecture that make it suitable for the real-time implementation of a large-scale system are composability and scalability. In a composable architecture, the properties of the main system follow its subsystems' properties. In real-time systems, the communication interface between the host computer in a node and the communication network is fully required in both the value domain and time domain. Scalable architecture requires the unlimited extensibility of the distributed system. In other words, the complexity of system operation should be independent of the system size.

In this chapter, we start with an overview of distributed computing platforms for real-time system architecture. The state messages facilitate the exchange of state information among the interconnected nodes within a communication network and enforce the autonomy of the nodes [151].

7.1 Introduction to parallel computing

Amongst the many benefits of parallel computing in dealing with large-scale and big-data analysis problems, one is to save time by providing superior time solution than traditional computing approaches (serial computing) and, thus, better completion. Further to this, solving larger/complex problems is no longer impossible only when the system is based on a single computer with limited computer memory. The approach of parallel computing using novel parallel algorithms is adopted [105]. To

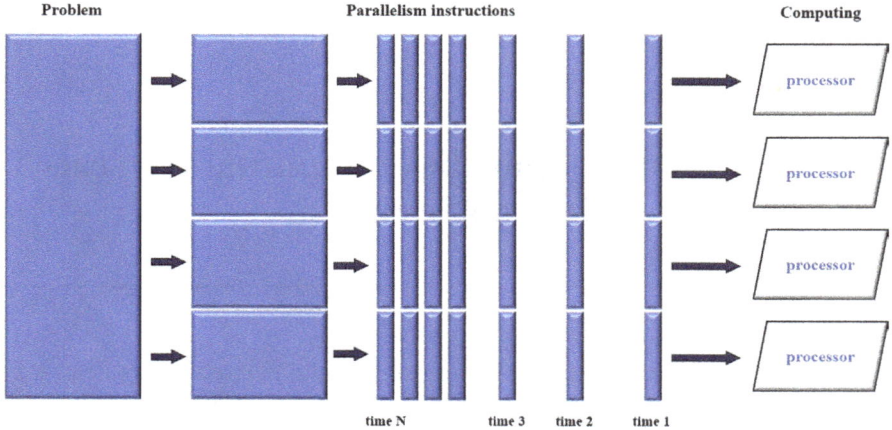

Figure 7.1   The proposed parallel approach with multiple-processor computing

explain this intuition, future computers which are parallel in architecture with multiple processors (multiple cores) can process many tasks concurrently and, thus are much more effective than classic ones as shown in Figure 7.1. Importantly, multiple processors in a computer can independently perform many tasks, which makes it possible to implement asynchronous parallel algorithms.

Moreover, high-level programming languages (i.e., MATLAB®, Python, Julia, R) and hardware nowadays also support parallel design and exploitation so as to facilitate the deployment of parallel algorithms, which are intended for parallel hardware and distributed architectures.

Disadvantages of parallel computing in optimisation

Besides the benefits, there are several critical issues in the use of parallel computing in optimisation [152]. Measure complexity and parallelisation of iterative methods are often difficult to implement. If this process is not chosen carefully, the parallel approach might not be efficient for reducing the solving time. Moreover, parallelism architectures for algorithm design will be extremely hard in stochastic optimisation problems since the full information of the problem is not available. Thus, how to choose communication aspects should be considered in parallel and distributed algorithms. In addition, synchronisation is very challenging in the development of parallel and distributed algorithms since any conflict in parallel algorithm processes might lead to poor performance.

To overcome these issues, each step of the algorithm process should be considered and implemented carefully by learning based on the training set of historical information of the problem. Hence, applying machine learning to parallel optimisation algorithms is a potential solution.

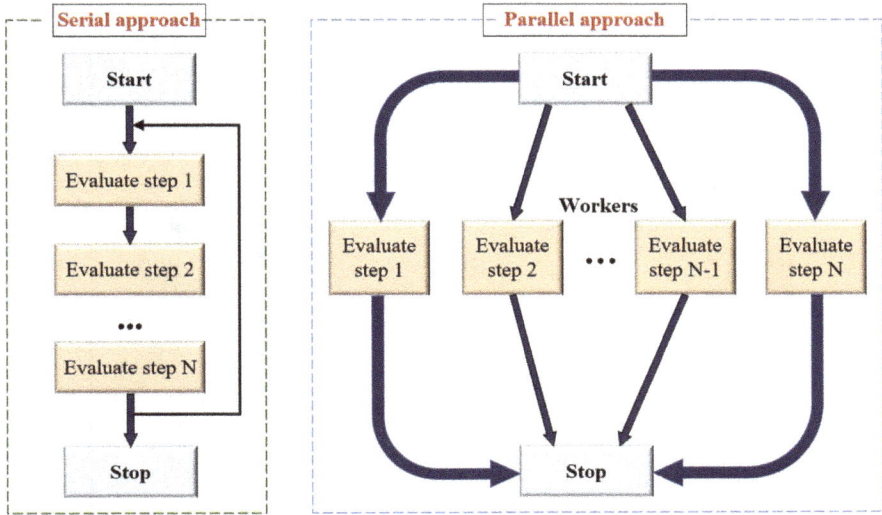

Figure 7.2   An example algorithm with parallel computing

7.2   The role of parallel computing in optimisation

A challenge for the use of parallel computing in optimisation is that as optimisation problems become very large-scale and expensive, solving them is also costly, i.e. large-size problems with massive numbers of variables or multiple-objective problems come with high expense in evaluating objective functions and constraints, implementing iteration and difficulty in guaranteeing the convergence rate. Existing optimisation methods might be prohibited through strict time deadlines such as real-time constraints because of the enormous and expensive cost of problem-solving, thus, calling for a faster and cost-saving solution. A focus should be on distributed and parallel computing [105, 153], which exploits multi-processor for operating multiple-task through the use of 'parallel optimisation algorithms', as shown in Figure 7.2.

In [153], parallel approaches were discussed to tackle several real-world optimisation problems. In the first group of problems, with a large number of variables and constraints (>100), a huge linear algebra calculation was required at each iteration, making gradient descent methods expensive. Parallel algorithms will speed up linear algebra evaluation and operate evaluation of constraints or derivative components simultaneously. The execution time of optimisation algorithms can be significantly reduced as in Figure 7.3, if we have an efficient parallel algorithm.

In the second group of problems, complex problems are formed by complicated objective functions and constraints which are expensive to evaluate. Using parallelism, individual complex function will be performed independently and concurrently, i.e., partial differential equations (subgradient methods). The third group of real-world optimisation problems includes the ones of slow convergence rate (i.e.,

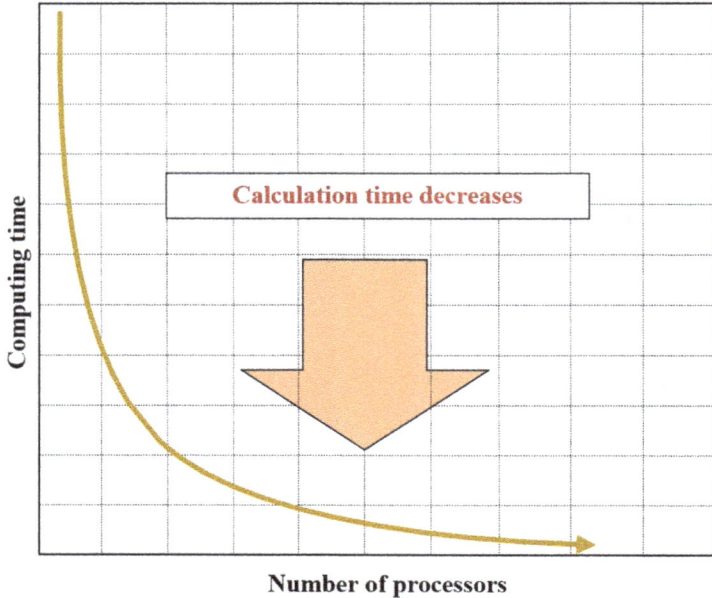

Figure 7.3   *The ideal benefit of parallel algorithm in reducing solving time*

requiring many iterations for convergence to a global solution). The iteration operation and function evaluation should be paralleled and performed concurrently to avoid this. Of course, in parallel computing, there may be multi-local optima existing in the solution. In this case, the final solution, which is frequently the best, is found by choosing the best solution among the local optima.

As the last group of the modern optimisation problems, general cases known as 'multi-objective problems' [154–158] consist of multi-objective problems and large-scale and complex problems. These problems require enormous linear algebra calculation and expensive derivative component operation. The parallelism approaches for these problems need to exploit many techniques, which have been introduced in the other groups. The combination of stochastic optimisation, e.g., genetic algorithms, evolutionary algorithms and parallel computing techniques [155, 159], is an efficient approach for multi-objective optimisation.

7.3   Parallel first-order optimisation approaches

To handle the massive computational and storage resources demanded by big data at reasonable power costs, we should rely on parallel and distributed computation. As discussed earlier, first-order methods are ideal for distributed and parallel computation [94]. A good communication between computers and local memory may enhance the overall numerical efficiency of synchronisation algorithms. For a synchronisation algorithm, first-order methods must coordinate the activities of computers on vector variables at each iteration. A conflict occurs when the single process takes an action much longer than the others. To alleviate the synchronisation problem, asynchronous algorithms allow updates using outdated versions of their parameters.

To describe the parallel first-order algorithm as in [160], we consider the first-order approximation via L-Lipschitz continuity as in (6.15) without $g(x)$. The next iteration is given as

$$x^{(k+1)} = \underset{x}{\arg\min} \left\{ \tilde{f}(x|x^{(k)}) \right\} \tag{7.1}$$

where $x \in \mathbb{R}^n$, $L > 0$ is the Lipschitz parameter and $\tilde{f}(x|x^{(k)}) = f(x^{(k)}) + \langle \nabla f(x^{(k)}), x - x^{(k)} \rangle + \frac{L}{2}\|x - x^{(k)}\|^2$.

Let us split the vector variable into blocks by a finite number of agents $1, ..., M$ All the agents update their block in parallel

$$x = \begin{bmatrix} x_1 \\ ... \\ x_M \end{bmatrix} \tag{7.2}$$

where x_m is controlled by agent m. Then, obtaining the next iteration in parallel approach

$$\hat{x}_m(x^{(k)}) = \underset{x_m}{\arg\min} \tilde{f}(x_m|x^{(k)}), \forall m = 1, .., M \tag{7.3}$$

The updated step of iteration is as

$$x_m^{(k+1)} = x_m^{(k)} + \beta^{(k)}(\hat{x}_m(x^{(k)}) - x_m^{(k)}), \forall m = 1, .., M \tag{7.4}$$

where $\beta^{(k)} \in (0, 1]$ is the stepsize.

The asynchronous algorithm also splits the vector variable into blocks by a finite number of agents $1, ..., M$. However, we need to break the synchronous scheme such that all agents continuously update their block in parallel without the coordinating mechanism in the central department by using the delayed time $x^{(k-d^k)}$ at each iteration as shown in Algorithm 6.

Simply put, the system processes each $f(.)$ of n objectives with one of m computers (processors). Thus, each machine stores data of algorithm corresponding to $\mathcal{O}(n/m)$-data samples since each $f(.)$ directly corresponds to a data point. Each processor communicates to the central unit to form the final gradient and achieves the ideal linear speed-up [94].

Algorithm 6 First-order asynchronous algorithm

1: **Select** a block m.
2: **Compute** $\hat{x}_m(x^{(k-d^k)}) = \underset{x_m}{\arg\min} \tilde{f}(x_m|x^{(k-d^k)})$.
3: **Update** block m: $x_m^{(k+1)} = x_m^{(k)} + \beta^{(k)}(\hat{x}_m(x^{(k-d^k)}) - x_m^{(k)})$.

Parallel embedded programming for real-time optimisation

We will discuss the parallel embedded system for real-time programming. This parallel computing approach is beneficial for complicated and large-scale problems. For $x = [x_1 ... x_M]^T$ controlled by a finite number of agents $1, ..., M$, supposing that $x^{(\kappa)} = [x_m^{(\kappa)}]_{m=1}^M$ is the feasible point at the κth iteration of problem (1.5). We sequentially address the function of $f(x)$ by iterating $(x_m^{(\kappa+1)})$. At the κth iteration, the following convex program is solved to generate the next feasible point $(x_m^{(\kappa+1)})$

$$\underset{x \in \chi}{\text{minimise}} \quad f^{(\kappa)}(x_{-m}) \tag{7.5a}$$

$$\text{s.t.} \quad g_m(x) \leq 0, \ m = 1, ..., M, \tag{7.5b}$$

$$h_p(x) = 0, \ p = 1, ..., P, \tag{7.5c}$$

where $f^{(\kappa)}(x_{-m}) = f(x_1^{(\kappa)}, ..., x_{m-1}^{(\kappa)}, x_m, x_{m+1}^{(\kappa)}, ..., x_M^{(\kappa)})$.

At the next iteration $\kappa + 1$, we define $x_m^{(\kappa+1)} = x_m^*$ where x_m^* is found as the optimal solution of problem (7.5). Hence,

$$(x_1^{(\kappa)}, ..., x_{m-1}^{(\kappa)}, x_m^{(\kappa+1)}, x_{m+1}^{(\kappa)}, ... x_M^k) \tag{7.6}$$

is a better feasible point than

$$(x_1^{(\kappa)}, ..., x_{m-1}^{(\kappa)}, x_m^{(\kappa)}, x_{m+1}^{(\kappa)}, x_M^k) \tag{7.7}$$

such that

$$f(x_1^{(\kappa)}, ..., x_{m-1}^{(\kappa)}, x_m^{(\kappa+1)} x_{m+1}^{(\kappa)}, ..., x_M^k) > f(x_1^{(\kappa)}, ..., x_{m-1}^{(\kappa)}, x_m^{(\kappa)}, x_{m+1}^{(\kappa)}, ..., x_M^k) \tag{7.8}$$

Thus, in Algorithm 7 we propose a parallel computational procedure for solving the large-scale optimisation problem (1.5).

The system processes each x_m of M variables with one of p computers (processors). Thus, each processor stores data of the algorithm corresponding to $\mathcal{O}(M/P)$-data samples since each $f(x_{-m})$ directly corresponds to a data point. Each processor communicates to the central unit to synchronise its iterative sub-solution for each point (x_m) and then achieves the speed-up from Algorithm 7.

Algorithm 7 Parallel computing for embedded optimisation programming

---

1: **Initialisation:**   Set $\kappa := 0$.   We separate a feasible point $(x_1^{(\kappa)}, x_2^{(\kappa)}, ..., x_M^{(\kappa)})$ for (1.5) into the sub-variables $(x_{M_1}, x_{M_2}, ..., x_{Mp})$ by calculating of $M_p = M/P$, $\forall p = 1, ..., P$, M_p is an integer and P is the number of processors.
2: **Repeat**
3:    **For** $p = 1, ..., P$ **do** (in parallel)
4:       **For** $i_p = 1, ..., M_p$ **do** (in sequence)
5:          Solve problem (7.2) for its optimal solution $(x_{i_p}^*)$ using automatic code generation in embedded optimisation programming.
6:    Set $\kappa := \kappa + 1$.
7:    Set $x_{i_p}^{(\kappa+1)} = x_{i_p}^*$, $i_p = 1, ..., M_p$, $p = 1, ..., P$.
8: **Until** convergence of problem (1.5).

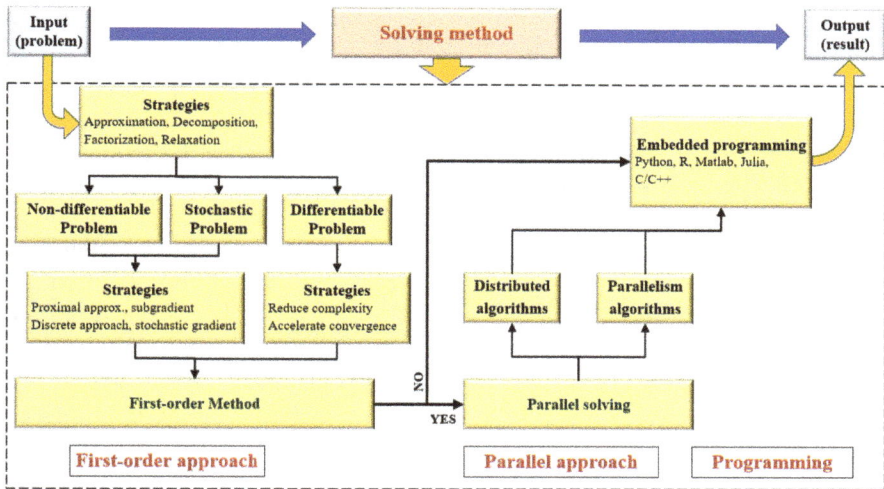

Figure 7.4   The proposed embedded optimisation system with first-order and parallel approaches

With the advantages of first-order and parallel approaches, the combination of the aforementioned methods will be more beneficial to large-scale optimisation problems in real-time applications. Figure 7.4 gives an illustration that corroborates the first-order and parallel/distributed methods in embedded systems.

7.4   Alternating direction method of multipliers

Another distributed computing approach, the alternating direction method of multipliers (ADMM), is an optimisation algorithm that separates an optimisation problem into smaller subproblems. These subproblems are often easier to solve and parallelise than the original problem [161].

The general optimisation form using ADMM is given by

$$\underset{x,z}{\text{minimize}} \quad f(x) + g(z) \tag{7.9a}$$

$$\text{s.t.} \quad Ax + Bz = c, \tag{7.9b}$$

$$x \in \mathcal{X}, z \in \mathcal{Z}, \tag{7.9c}$$

where f and g are convex functions and $\mathcal{X}$ and $\mathcal{Z}$ are the domain sets of variables x and z.

The augmented Lagrangian is as follows:

$$\mathcal{L}_\rho(x, y, z) = f(x) + g(z) + y^T(Ax + Bz - c) + \frac{\rho}{2}\|Ax + Bz - c\|_2^2 \tag{7.10}$$

Then the general ADMM algorithm based on the updated (x, z, y) is given by

$$x^{k+1} := \text{argmin}_x \mathcal{L}_\rho(x, z^k, y^k) \tag{7.11a}$$

$$z^{k+1} := \text{argmin}_z \mathcal{L}_\rho(x^{k+1}, z, y^k) \tag{7.11b}$$

$$y^{k+1} := y^k + \rho(Ax^{k+1} + Bz^{k+1} - c). \tag{7.11c}$$

The convergence of ADMM algorithm will be achieved when the feasible iteration of $Ax^k + Bz^k - c \to 0$ or the objective function approaches the optimal value as $f(x^k) + g(z^k) \to p^*$.

Combining linear and quadratic terms in augmented Lagrangian method, we have

$$\mathcal{L}_\rho(x, y, z) = f(x) + g(z) + y^T(Ax + Bz - c) + \frac{\rho}{2}\|Ax + Bz - c\|_2^2 = f(x) + g(z) + \frac{\rho}{2}\|Ax +$$
$$Bz - c + u\|_2^2 + \text{const.}$$
$$\tag{7.12}$$

where $u = y/\rho$. The ADMM iteration based on the updated (x, z, u) with $u^k = y^k/\rho$ is given as

$$x^{k+1} := \text{argmin}_x \left\{ f(x) + \frac{\rho}{2}\|Ax + bz^k - c + u^k\|_2^2 \right\} \tag{7.13a}$$

$$z^{k+1} := \text{argmin}_z \left\{ g(z) + \frac{\rho}{2}\|Ax^{k+1} + bz - c + u^k\|_2^2 \right\} \tag{7.13b}$$

$$u^{k+1} := u^k + (Ax^{k+1} + Bz^{k+1} - c). \tag{7.13c}$$

An example of ADMM algorithm as in [161] considers the following lasso problem

$$\frac{1}{2}\|Ax - b\|_2^2 + \lambda\|x\|_1 \tag{7.14}$$

Transforming problem (7.14) into ADMM form results in

$$\frac{1}{2}\|Ax - b\|_2^2 + \lambda\|z\|_1 \text{ s.t. } x-z = 0 \tag{7.15}$$

Then, the ADMM algorithm is as follows:

$$x^{k+1} := (A^T A + \rho I)^{-1}(A^T b + \rho z^k - y^k) \tag{7.16}$$

$$z^{k+1} := S_{\lambda/\rho}(x^{k+1} + y^k/\rho). \tag{7.17}$$

$$u^{k+1} := y^k + \rho(x^{k+1} - z^{k+1}) \tag{7.18}$$

where $S_\alpha(x) = (1 - \alpha/\|x\|_2)_+ x$ is the vector soft thresholding operator.

Two efficient approaches, known as consensus and splitting optimisation via ADMM, can deal with some challenges in practical real-time multi-objective problems. They are fast and reliable while their implementation is not difficult with free code library. More interestingly, these schemes can easily take in parallelism to solve large-scale problems. The detail of these approaches can be seen in [161, 162]. ADMM algorithms are efficient for simple large-scale convex problems since the fundamental of ADMM algorithms is to solve a problem by separating it into smaller problems as in distributed approach. And thus, the evaluation cost of each iteration is inexpensive.

The application of ADMM algorithms in parallel approaches can deal with many complex large-scale problems [94, 163, 164]. For a consensus problem as shown in [164], the variable splitting, Jacobian-type algorithm and proximal Jacobian via ADMM technique are appropriate for distributing operation and attractive for solving large-scale problems. In the first method, this distributed ADMM approach introduces splitting variables when the problem substantially increases the number of variables and constraints. In the second method, Jacobian-Gauss-Seidel ADMM algorithm replaces two-block alternating minimisation schemes by a sweep of Gauss-Seidel update in parallel computing. In the last method, proximal Jacobian ADMM can provide global convergence as well as a convergence rate of $\mathcal{O}(1/k)$. More importantly, this scheme can deal with non-smooth problems by replacing non-smooth functions of the problem with their proximal term. In such implementation, an open-source optimisation tool of parADMM is proposed for parallelism ADMM method in [165]. parADMM can exploit ADMM step operations with the distributed resource using a graphic processing unit (GPU) or multiple CPU cores. In fact, parADMM solvers allow parallelism in shared-memory multiple processors and use multiple independent machines concurrently.

However, ADMM algorithm may be inefficient for complex problems with costly evaluation at each iteration, constrained problems or non-smooth problems [94, 163].

7.5   Applications of parallel computing in 5G Internet of Things

Parallel computing has been widely used in large-scale stochastic optimisation problems [166–168] with big data analysis [169].

The approach of parallelism has been applied to solve antenna design problems. For instance, a dual-band (1.9 and 2.4 GHz) antenna element is designed using electromagnetic genetic optimisation for wireless communication applications. To this end, an electromagnetic genetic optimisation algorithm is developed for the cluster supercomputing platform [170], allowing users to combine the accuracy of electromagnetic genetic optimisation analysis and the speed of a parallel computing algorithm. In another study [171], multi-band and wide-band patch antenna design are developed by a novel evolutionary optimisation method, which is the combination of particle swarm optimisation (PSO) and finite-difference time-domain (FDTD) to achieve the optimal antenna solution under a design standard. PSO optimises the antenna parameters while FDTD simulators evaluate a fitness function to represent the candidate design performance. Using parallelism clusters, the computational time is significantly reduced for optimisation process. Rectangular and E-shaped patch antennas are two examples in [171] for testing the parallel PSO/FDTD algorithm.

Parallelism scheme is promising in mobile applications since it reduces the execution time and energy consumption in the growing application processing loads associated with the explosion of mobile networks and complex applications.

A simple framework, called ThinkAir, is proposed to migrate mobile networks into the cloud [172]. By using ThinkAir, virtualised mobile network concept in method-level computation offloading will enable dynamic resource allocation. Furthermore, mobile cloud may scale up the network and enhance the power of cloud computing by executing parallelism computing using multiple virtual machine (VM) images. In particular, using multiprocessor in mobile cloud computing, a parallelism execution splits the tasks of network among multiple VMs. There are two kinds of parallelisation algorithms, namely recursive algorithms and massive-data algorithms. The former parallelise sub-solution computations based on a solution construct, while the latter split computations based on the analysis of data intervals.

Decomposability concept is a key for resource allocation in network utility maximisation. It provides an appropriate distributed algorithm for the resource allocation problem of a given network. A tutorial on decomposition methods is shown in [173] in the mathematical language of decomposition theory to develop modularised and distributed control of network design. Distributed solutions will be efficient in large-scale networks while the centralised solution is too complex, too expensive or infeasible. In the context of distributed theory, many methods are provided and exploited such as distributed subgradient method or Jacobi and Gauss-Seidel parallel iterations. For instance, a problem is decomposed into several subproblems controlled by a master problem through prices (dual decomposition) or direct resource allocation (primal decomposition). The master problem and subproblems communicate through a form of message passing in the network overhead. Similarly, a parallel resource allocation scheme for multi-radio access in heterogeneous wireless networks is introduced in [174]. Therein, a mobile station can simultaneously transmit over multiple radio access technologies. Then, a distributed joint allocation algorithm is provided for maximising the network capacity. For wireless sensor networks, the decentralised detection and estimation in wireless sensor applications have been considered in [175]. Sensor networks are often distributed with different sensors accessing the central system under various data streams. This indicates that one sensor might not share all data with the central centre or other sensors. For a universal network consistency, a larger learning problem is considered with only a small portion of that allocated to each sensor and handled by distributed learning within communication constraints.

Parallel and distributed simulation technologies are introduced in [176] and the references therein. Wireless mobile and ad hoc networks will enormously expand, leading to large and complex system design and, thus, to many critical challenges for parallel and distributed discrete-event simulation and implementation. Therefore, novel parallel and distributed methods need to be robust and dynamic. Parallel structures of the simulation are to ensure strict time for large-scale simulation where multiple events are processed at one time. In a parallel simulator, a set of logical processes (LPs) interacts through messages, each of which carries an event and its timestamp, called event message. A subset of the model state is managed by each LP (local state). In the local state, an LP receives the event which represents a transition. Finally, the simulation of events schedules the events as event messages and transfers to neighbouring LPs for the next process. In a simulation, events must always be executed in increasing order. However, when an event is incorrectly processed before the simulated time, which affects state parameters in the subsequent events, this may lead to a causality error in which the future events could influence the present events. To deal with this problem, some synchronisation protocols should be exploited such as conservative and optimistic schemes. Conservative synchronisation techniques use blocking process to avoid violation of constraints. Optimistic methods detect the synchronisation errors at run time and recover the event at the process before the violation happened using a recovery mechanism.

A library of parallel network simulators is designed in [177], called GloMoSim (global mobile system simulator), for parallel simulation of wireless networks. Therein, communication network protocols are separated into a set of layers with its application programming interface (API). In each layer, models of protocols cooperate with a lower or higher layer via API, art and these models are compared by implementing different communication standards and modulation schemes. Then, synchronisation protocols perform the GloMoSim parallel implementation.

Chapter 8

Machine learning for real-time optimisation

With the boundless quality-of-service (QoS) requirements of 5G networks and beyond, communication systems must be more dynamic and intelligent and satisfy many network demands simultaneously. The concepts of artificial intelligence (AI), machine learning (ML), deep learning (DL) [178–181], as depicted in Figure 8.1, have found their applications in wireless communications [106–110, 182, 183].

As wireless communication systems in dynamic environments rapidly change over time, more unexpected behaviour patterns and complicated scenarios will develop. Fortunately, ML can use robust algorithms to calibrate itself to newly acquired knowledge, provide low-complexity estimates for system model, support self-organising systems with limited human intervention. For instance, in resource allocation optimisation problems, ML attempts to reduce the complexity of optimisation problems by shrinking the solution space using feature selection technique and employing meta-heuristic solution methods for multi-objective optimisation such as finding initial solutions or choosing appropriate heuristic methods.

In this book, we assume that the reader is familiar with the basics of ML. The reader may also refer to some very excellent books in [184-188] and their references. Now, we move on to the ML application for 5G-IoT (Internet of Things) in real-time optimisation context.

8.1 Brief overview of machine learning for wireless communication

Wireless communication networks need to operate in an adaptable, flexible and intelligent manner – this leads to ML being deployed in self-organising networks (SONs). A SON will adapt, autonomously scale and control the network in a stable stage and be agile enough to maintain the network requirements [189–192]. A typical SON's operation consists of self-configuration, self-optimisation and self-healing as shown in Figure 8.2. Self-configuration performs the auto-configuration process of the network operation in cellular base stations such as IP configuration, policies and radio resource. Self-configuration is useful in controlling and expanding the network, i.e., when the base stations are controlled for activating or changing the network parameters by themselves to improve the total network performance. A self-optimisation scheme is exploited for optimising network parameters in order to guarantee a better

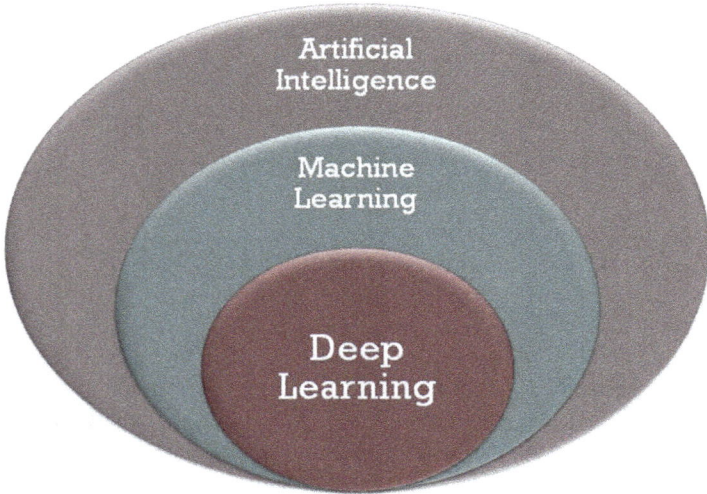

Figure 8.1   From AI to ML and DL. AI involves machines performing tasks that exactly mimic characteristics of human intelligence. ML, introduced in the 1950s, is a pathway to achieving AI. DL is a subset of ML and entails methods such as deep artificial neural networks (ANNs), deep reinforcement learning (RL).

network performance. There are many optimisation problems such as beamforming design, interference alignment, fronthaul and backhaul optimisation, resource allocation (sum rate, energy efficiency), antenna parameters, or mobility and load balancing. By real-time monitoring the feedback from the central control system, self-optimisation schemes optimise and maintain the network performance. For large-scale networks, distributed self-configuration and self-optimisation schemes are exploited with the help of self-healing functions. Since the full information of any system is imperfect and errors and network issues might occur suddenly out of control, a self-healing scheme will be triggered to continuously monitor the system in order to ensure a fast recovery.

Some surveys of SONs using ML techniques are shown in [109, 193]. In general, ML methods can be classified into two major categories, namely supervised and unsupervised learning. Supervised learning requires full information of input and output system for the training process. In contrast, unsupervised learning does not have a supervisor, the expected output is unknown, and thus, the system needs to learn by itself. From the survey in [109], a block diagram of common ML methods in the context of SONs is depicted in Figure 8.3.

These characteristics of ML methods have also been analysed and applied for appropriate problems in SONs [109, 193] as shown in Figure 8.4. For instance, genetic algorithms (GAs) can be employed for spectrum and parameters optimisation; ANNs are used for learning and adapting to complex environments and improving spectrum resources; reinforcement learning (RL) is to maximise spectrum

Figure 8.2   SONs using ML

utilisation in the long term; support vector machine is utilised for channel selection, transmission parameters adaption and beamforming design.

Some surveys of cognitive radio networks (CRNs) using ML techniques are shown in [193–195]. By awareness and understanding of the environment, a CRN can learn and adapt to statistical environmental variations for simulation in assessment performance and deployment in reality [196]. CRNs should exploit the cognitive engine's abilities to learn and reason by making use of ML algorithms. When system information (inputs and outputs) is unknown, unsupervised learning becomes necessary. For example, to deal with imperfect wireless channel estimates, which may lead to an increase in probability error, ML methods can be applied to yield the wireless channel estimation.

Perception, learning and reasoning are three processes of ML in CRNs [194]. Perception is the ability to understand the surrounding environment to acquire information. By using classification methods and hypothesis generalisation, learning is to transform the acquired information into learning knowledge. Then, reasoning uses this knowledge to achieve system targets. In [195], a concept of innovative cognitive radio systems is proposed, known as COgnition-BAsed NETworkS (COBANETS). This approach uses advanced ML methods, i.e., unsupervised DL and probabilistic generative models, to develop system-wide learning, modelling and optimisation applications. As another aspect of cognitive radio with ML, cognitive cycle is developed by the learning process with the sequence of sensing, observing, learning and

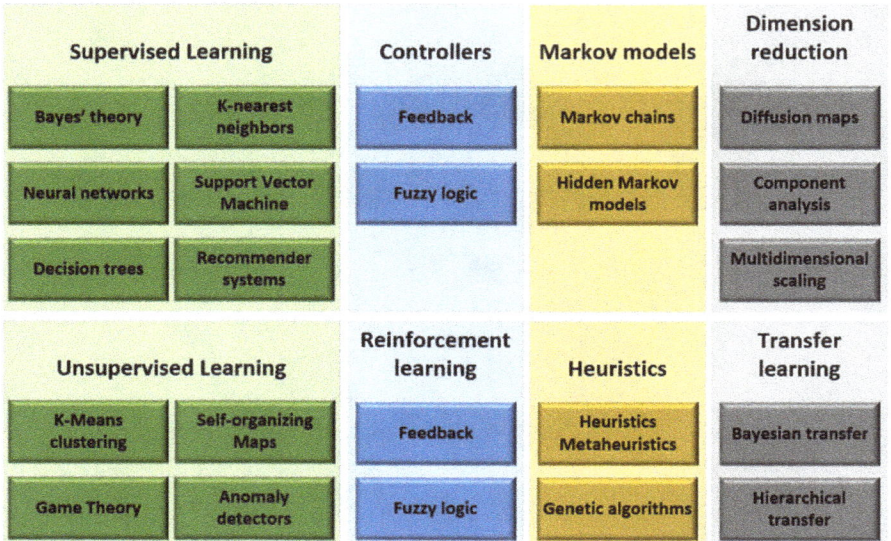

Figure 8.3   A block diagram of common machine learning methods in SONs (Klaine et.al. 2017).

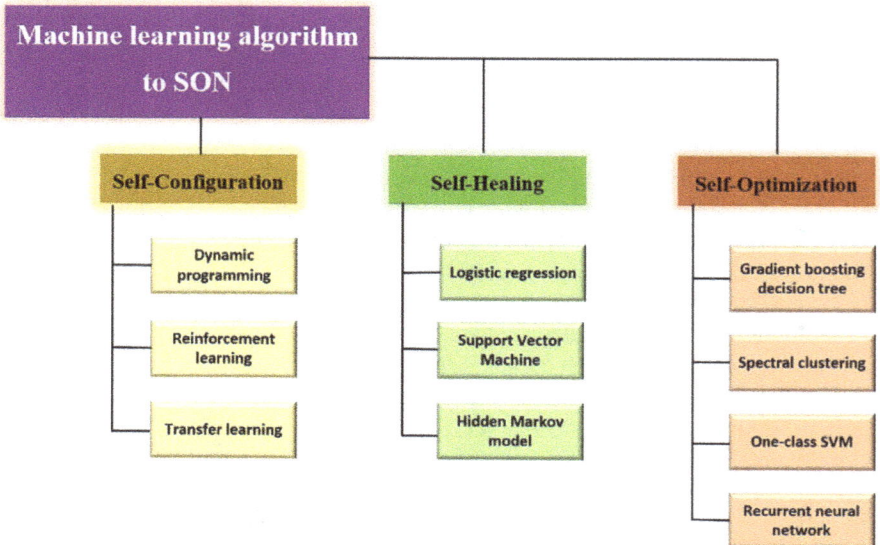

Figure 8.4   Self-organized networks using machine learning (Latif el. al, 2017).

making decision [193]. Support vector machine (SVM), ANN, meta-heuristic algorithms, fuzzy logic, Markov models, Bayesian learning and RL are some popular ML approaches for cognitive cycle.

In wireless sensor networks (WSNs), ML is an efficient approach to avoid unnecessary redesign in unstable wireless sensor systems [107, 197, 198]. An intelligent sensor network has to adapt its behaviour for optimising the ability to gather real data from communication systems [197]. An intelligent sensor system may have functions such as self-calibration, self-validation and self-compensation. Self-calibration allows the sensors to supervise the measurement in order to make the decision to trigger a new calibration. Self-validation uses mathematical modelling to estimate error propagation and isolation error. Then, a high performance is achieved by using compensation methods in self-compensation function. There are many challenges of WSNs that require the application of ML, e.g., scheduling and routing, clustering and data aggregation, event detection and query processing, localisation and object targeting, security and anomaly intrusion detection, data integrity and fault detection and resource management [107,198].

The next-generation wireless networks are very complex systems due to the enormous service requirements and massive data rate, various applications, devices and networks. Unlike traditional networking approaches, e.g., data analysis, error probability analysis and optimisation models, which might be inappropriate for modern wireless communication with large-scale problems and massive data analysis [182], ML is a promising AI tool to support smart radio networks [106,108,110,182,183,199]. As a result, ML can be widely used in the paradigm of next-generation systems, such as massive multiple-input multiple-output (MIMO), device-to-device (D2D) networks, HetNets, CRNs and so on, as shown in Figure 8.5 [108].

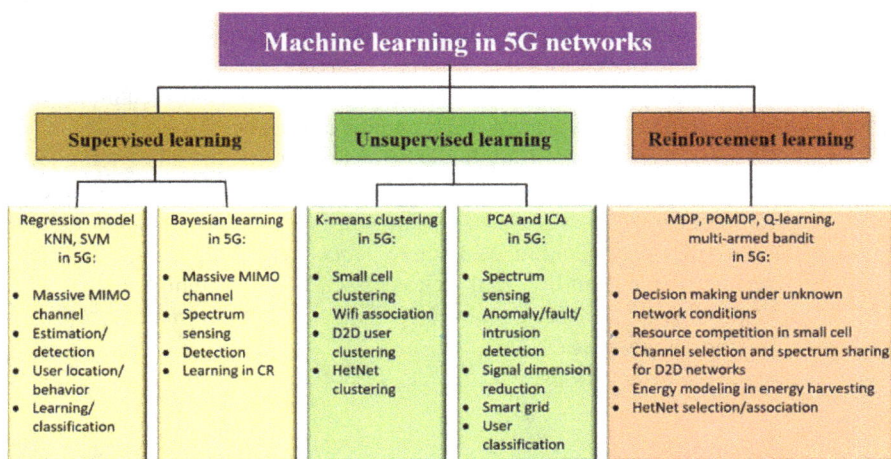

Figure 8.5   A diagram of ML in 5G networks

8.2 Interplay of machine learning and optimisation

The interplay between ML and optimisation programming results in two major directions [200]: applying optimisation to ML and applying ML to optimisation.

Optimisation problems have been considered as the heart of most ML approaches [201–205]. This is because many ML problems can be reduced to optimisation problems, or the existing optimisation methods are used for supporting ML with new learning models. By solving a core optimisation problem, the variables or parameters with respect to the loss functions or regularisation functions in ML models can be optimised. However, optimisation problems, which exist in ML, are often rapidly changing in time. Based on a specific learning scheme, the adaptation and validation of core optimisation processes are necessary to advance research in ML.

On the other hand, modern optimisation has been efficiently utilising existing ML methods [206–212]. Most classes of optimisation programming allow the integration of ML methods to develop new optimisation techniques [200]. As large-scale problems arise, existing optimisation algorithms might be infeasible. The collaboration of ML and optimisation techniques offers simple and efficient algorithms which are promising in dealing with modern optimisation problems in real-time applications under reasonable computational time. Optimisation methods that exploit ML models can handle most non-deterministic polynomial-time hardness (NP-hardness) problems and more importantly, it is an approach to real-time applications. SVM [203, 213, 214], deep neural networks (DNNs) [178–181], RL [215, 216] are efficient ML techniques that can be applied in novel optimisation methods. A hybrid method consisting of ML and optimisation can be tailored for real-life applications.

ML implies learning from computational procedure over successive iterations. It is used to build an approximation of the objective function and guide the good choice made in the next iteration. ML methods such as ANNs can estimate the outcome of a series of sub-problems all at once, attempt to reduce the complexity of a constrained continuous optimisation problem by shrinking the solution space using a feature selection technique, and adjust meta-heuristic solution methods for multi-objective optimisation, e.g, choosing appropriate heuristic methods or finding initial solutions. In fact, ML stands for the development of algorithms or techniques that learn from observed data by assembling mathematical models.

Desirable properties of optimisation algorithms using ML include good generalisation, exploitation of problem structure, scalability to large problems, good performance in practice in terms of execution time and memory requirements, simple and easy algorithm implementation, fast convergence to an approximate solution, robustness, and numerical stability for the class of ML models attempted.

In the next part, we introduce a novel optimisation algorithm by applying a popular DL method, known as DNNs.

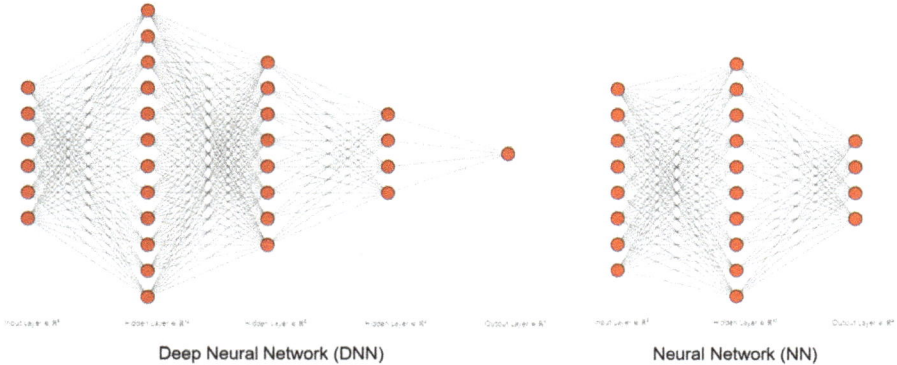

Deep Neural Network (DNN)   Neural Network (NN)

Figure 8.6   *An ANN consists of three types of layers, i.e., input, hidden and output layer. A simple neural network has one input and one output layer and one hidden layer in between, whereas a DNN has multiple hidden layers.*

8.3 Deep neural networks

A simple, fully connected neural network (or DNN) has three types of layers: one input layer, one (or multiple) hidden layer(s) and one output layer, as shown in Figure 8.6.

At the input layer, the input vector $x = (x_1, ..., x_{dx})$ is known, where d_x is the dimensional input vector or the number of input nodes. Information from d_x will be passed to all the nodes of the next layer after multiplication with associated weights. For a given node in the next layer, a signal is transformed by an activation function before it is continuously passed to the next layer(s) called 'hidden layer(s)' $h_j = (h_{j1}, ..., h_{jd_j})$, where j is the jth hidden layer and d_j is the number of nodes in jth hidden layer or output layer in the final path. At the last layer of the network, known as the output layer, the output vector values $p = (p_1, ..., p_{d_p})$ are predicted by the values passed from the previous layer(s) where d_p is the dimensional output layer or the number of output nodes. An example of a DNN is provided in Figure 8.7 with two hidden layers, $d_x = 5$, $d_p = 3$ and $d_j = 4$, $j = 1, 2$.

Activation functions are an important feature of ANNs [181, 217]. They decide whether a node in the layer, which is receiving information relevant to the given information, should be activated or not

$$Y = \text{activation}\left(\sum (\text{weight} * \text{input}) + \text{bias} \right) \qquad (8.1)$$

An activation function is a nonlinear transformation over the input signal which is then sent as input to the next layer. A neural network without an activation function is a linear regression model with limited ability to solve complex problems. Based on the characteristics of the problem the 'weights' and 'bias' would be chosen and adapted in activation functions for easy implementation and quicker convergence. There are several popular activation functions, as shown in Table 8.1.

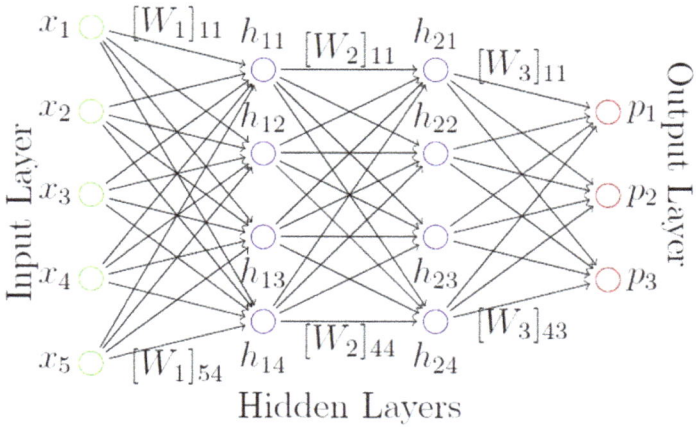

Figure 8.7   A fully connected DNN with two hidden layers

In [209], a ML-based approach is proposed for optimisation problems in wireless resources as shown in Figure 8.8. Therein, an optimisation algorithm will be trained and learn its input/output relation by using a DNN model [178, 211] at the training stage. In the testing stage, we do not need to implement the optimisation algorithm again but instead present the optimisation process as a 'black box'. Several network layers will approximate a training set of resource management algorithms by using a DNN model, which requires simple operations to implement a finite training sample set. With the sufficient training data set, we can reduce a lot of processing time for optimising wireless resource allocation in the testing stage.

However, existing optimisation algorithms for resource allocation might leave many challenges for the training stage (offline stage). Many optimisation algorithms have followed a flowchart of calculation processes and repeated to an expected achievement after a number of iterations. The achievement is often a saturation point in the domain of the objective function in the considered problem, as shown in Figure 8.9. At the beginning of the algorithm, the iteration initialises with a feasible point, then applies the appropriate method to find a feasible solution by computing a step vector operation. The updated formula is obtained for checking the convergence under a criterion.

Following Figure 8.9, we introduce Algorithm 8 to sum up the general optimisation algorithm.

Table 8.1   Some activation functions

Functions	Operations	Useful in
Identity	$f(x) = 1, x \geq 0$	Binary classifier
Binary step	$f(x) = ax$	Activating multiple nodes simultaneously
Sigmoid	$f(x) = \frac{1}{1+e^{-x}}$	Widely used for smooth functions
Tanh	$\tanh(x) = 2 \times \text{sigmoid}(2x) - 1 = \frac{2}{1+e^{-2x}} - 1$	Widely used in many cases
ReLU	$f(x) = \max(0, x)$	Activating all nodes simultaneously
Leaky ReLU	$f(x) = \begin{cases} ax, x<0 \\ x, x \geq 0 \end{cases}$	Improve ReLU
Softmax	$f_j(x) = \frac{e^{x_j}}{\sum\limits_{K=1} e^{x_k}}, j = 1, ..., K$	Used in the output layer of classifier

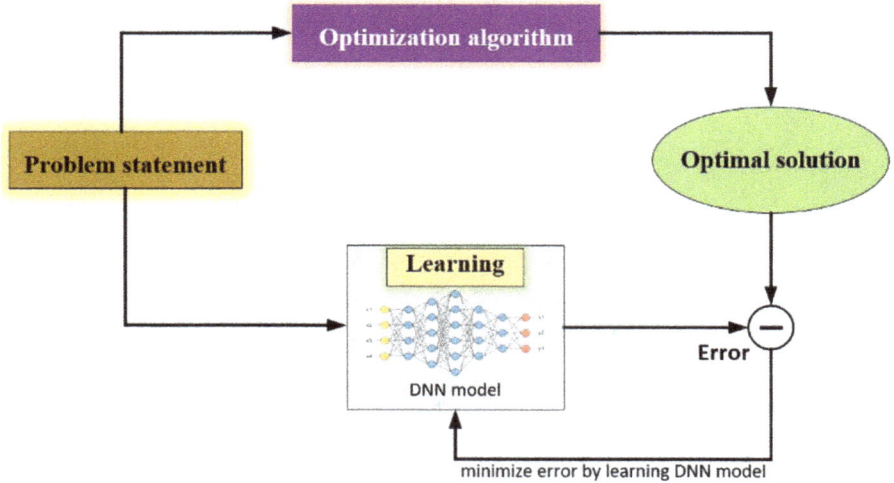

Figure 8.8   *A DNN-based approach to optimisation [209]. At the training stage, a DNN model is used to learn resource allocation optimisation algorithm. At the testing stage, the resource allocation is performed by the DNN model instead of the optimisation algorithm. ©IEEE 2018. Reprinted with permission from [218].*

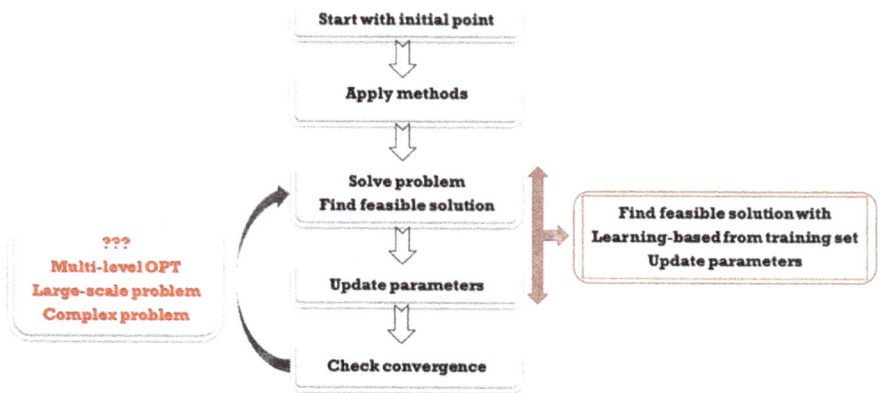

Figure 8.9   *A flowchart of general optimisation algorithm*

Algorithm 8 General optimisation algorithm

1: **Input** optimisation problems (variables, objective function, constraints).
2: **Output** Optimal variables, optimal performance of objective function.
3: **Initialisation** Set $k = 0$. Choose the feasible point of $x^{(0)}$.
4: **Repeat:**
5: $\triangle x := \psi(x^{(k)}, f(x^{(k)}, \nabla f(x^{(k)})))$
6: Set $k = k + 1$
7: Generate the next point $x^{(k+1)} = x^{(k)} + \triangle x$
8: **Stop:** the convergence is satisfied.

For instance, $\psi()$ represents the applied method, i.e., classic gradient descent $\psi(.) = -\beta \nabla f(x^{(k)})$ with β being stepsize, $x^{(k)}$th being the iterative point; momentum $\psi(.) = -\beta \left(\sum_{j=0}^{k-1} \alpha^{k-1-j} \nabla f(x^{(j)}) \right)$ and learning algorithm $\psi(.)$ as in a neural network.

Although the training stage may allow plenty of time for the implementation, it is difficult for a DNN model to deal with large-scale complex optimisation problems and a finite number of training samples. Furthermore, as the system's environment changes rapidly, the problem can change very fast, thus the training stage may not be in accordance with the new system environment. Hence, the optimisation algorithm design or the training stage of learning optimisation with DNNs has to become faster for adapting to the rapid changes of the system environment. A fast training stage or online training stage is an attractive approach for learning-based optimisation algorithm in real-time applications. An online training database platform is needed to support data analytics and applications and produce data-driven insights in a timely manner by allowing to access useful and aggregated data from many sources. Gathering and analysing database in parallelism, increasing variability of training dataset during the learning process and real-time augmentation in learning and classification datasets are potential approaches for online training stage of DNN models.

At the training stage in Figure 8.10, a learning-based optimisation algorithm is provided. The main goal of this scheme is three-fold, the first being to learn the optimisation algorithm for adapting the algorithm with real-time application. Second, as discussed above, some novel techniques in ML are efficient in reducing the solving time of the learning-based optimisation algorithm. Last but not least, the embedded optimisation system is also necessary for the implementation of real-time optimisation in the ML and optimisation context.

If the learning-based optimisation algorithm learns the updated formula, it can learn a new algorithm that is modelled as a neural network [208, 210, 219]. Learning the weights of the neural network and parameterising the updated formula of the algorithm can provide useful function approximators, model any updated formula with sufficient capacity, allow for efficient search and easily perform the training process with backpropagation. Therefore, the appropriate optimiser would simply

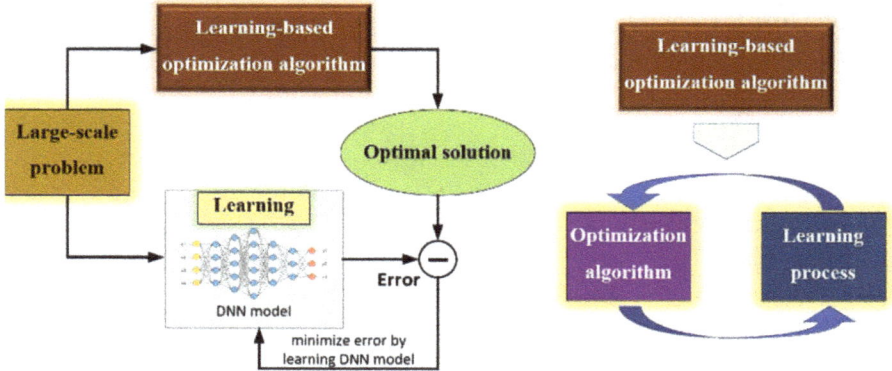

Figure 8.10   *A DNN-based approach to real-time optimisation. At the training stage, we try to learn the optimisation algorithm that can approach online training stage. Next, the testing stage is implemented in real-time application. ©IEEE 2018. Reprinted with permission from [218].*

memorise the optimum, and after learning with a sufficient training set, the optimiser then converges to the optimum within a few steps regardless of initialisation in the future.

In the next part, we introduce an efficient ML approach called RL, for learning the optimiser or optimisation algorithm.

8.4   Reinforcement learning

In a given environment, an agent which selects actions and receives feedback after each action is taken, interacts with this environment. This agent will predict how

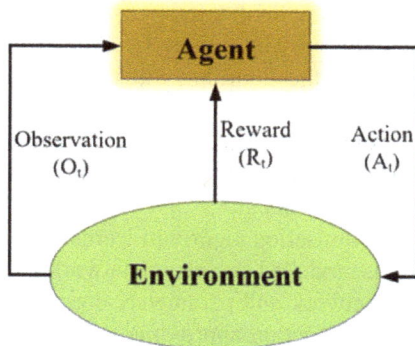

Figure 8.11   *A RL model*

good or bad the new state is. The major mission of RL is trying to pick actions for the agent based on the current state by choosing a good way [210].

Combining the RL model in Figure 8.11 and the optimisation algorithm in Algorithm 8, we can arrive at a learning-based optimisation algorithm with RL as in [210]. An agent is a primary component in the intelligent process of making decision by observing the statistics of the previous gradients and iterates. The environment is the goal for the agent to perform iterations by the action ($\triangle x$) for the next stage ($x^{(k+1)}$) at the performance cost ($f(x^{(k+1)})$). The action, which is controlled by a policy ($\psi(x^{(k)}, f(x^{(k)}), \nabla f(x^{(k)}))$), is used to update the iterate. By learning the policy, the updated formula is also learned. To train the agents, rewards are used.

Chapter 9
Real-time embedded convex programming

A diagram of an embedded convex optimisation system is given in Figure 9.1. As illustrated, a practical optimisation problem is the input of the system. Efficient methods are used for solving the problem on a computer (central processing unit (CPU), graphics processing unit (GPU), Field Programmable Gate Array (FPGA)), applying novel approaches for real-time optimisation, e.g., first-order methods, parallel approaches and learning-based optimisation algorithm. A custom code is produced via manual or automatic code generation by an embedded software before being integrated into the real system.

There are many optimisation software packages available for both research and practical use. In this chapter, we will introduce some free optimisation tools that can be exploited for real-time optimisation applications.

9.1 Real-time operating systems and programming languages for embedded systems

9.1.1 Complexity of convex optimisation problem

The computational complexity of a convex optimisation problem is calculated as

$$\mathcal{O}(N_{var}^2 N_{con}^{2.5} + N_{con}^{3.5}) \tag{9.1}$$

where N_{var} is the number of scalar variables and N_{con} is the number of constraints.

9.1.2 Running time of algorithms

An algorithm with rational input is said to run in polynomial time if there is an integer k such that it runs in $\mathcal{O}(n^k)$ time, where n is the input size, and all the numbers in intermediate computations can be stored with $\mathcal{O}(n^k)$ bits. An algorithm with arbitrary input is said to run in strongly polynomial time if there is an integer k such that it runs in $\mathcal{O}(n^k)$ time for any input consisting of n numbers and it runs in polynomial time for rational input. In the case $k = 1$ we have a linear-time algorithm. If a function is computed by some polynomial-time algorithm, it is said to be computable in polynomial time. Polynomial-time algorithms are sometimes called "good" or "efficient" [220].

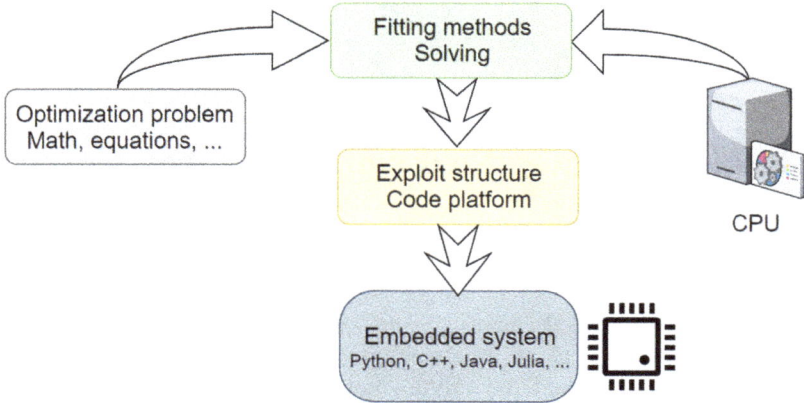

Figure 9.1   An embedded optimisation system

As Table 9.1 shows, polynomial-time algorithms are faster for sufficiently large instances. For various input sizes n, we show the running time of algorithms that take $\mathcal{O}(n^k)$, $10n^2$, $n^{3.5}$, $n^{\log n}$, 2^n and $n!$ elementary steps. The table also illustrates that constant factors of moderate size are not very important when considering the asymptotic growth of the running time. We assume that one elementary step takes one nanosecond. As illustrated in Figure 9.2, many optimisation problems we focus on in this book are of moderate sizes and to be solved in microseconds to seconds.

9.2   Embedded optimisation software

9.2.1   Embedded in MATLAB®

Two widely used optimisation tools compatible with MATLAB are CVX [221] and YALMIP [222]. CVX is a MATLAB-based tool that turns MATLAB into a modelling language. Therein, the constraints and objectives of an optimisation problem are determined by standard MATLAB expressions. In default mode, CVX provides a disciplined convex programming to approach convex optimisation. Convex functions and constraints are built up from the convex rules set in a base library. These rules address constraints and objectives following a canonical form after being automatically transformed and solved. It is not difficult to start CVX since it includes a growing library of examples from the book *Convex Optimisation* [1] and from the plenty of applications to help users.

YALMIP is also a MATLAB-based tool for working on two most important convex optimisation programs in control and systems theory, known as semidefinite programming (SDP) and linear matrix inequalities. YALMIP works with polynomials, function values, derivatives, integrals and their properties. For instance, to model and solve SDP problems, YALMIP toolbox simplifies the problems into

Table 9.1   Estimated running time for some algorithms

n	$100n \log n$	$10n^2$	$n^{3.5}$	$n^{\log n}$	2^n	$n!$
10	3 μs	1 μs	3 μs	2 μs	1 μs	4 μs
50	28 μs	25 μs	884 μs	4 s	13 d	NA
100	66 μs	100 μs	10 ms	5 h	4.10^{13} y	NA
1 000	1 ms	10 ms	32 s	3.10^{13} y	NA	NA
10^4	13 ms	1 s	28 h	NA	NA	NA
10^6	2 s	3 h	3 169 y	NA	NA	NA
10^8	266 s	3 y	3.10^{10} y	NA	NA	NA
10^{10}	9 h	3.10^4 y	NA	NA	NA	NA

general, oriented control cases and solve them in simple ways using external solvers. By standard MATLAB expressions, SDP algorithm prototypes can be implemented in time scales of minutes.

9.2.2   Embedded in Python programming

CVXPY [223] is a modelling language embedded in Python for convex optimisation problems. Python is a high-level programming language used in many general programmings with various purposes. Python programming emphasises code readability and provides constructs for clearly programming on both small- and large-scale problems. CVXPY can express optimisation problems in efficient ways following math models rather than restricted standard forms by solvers such as CVX

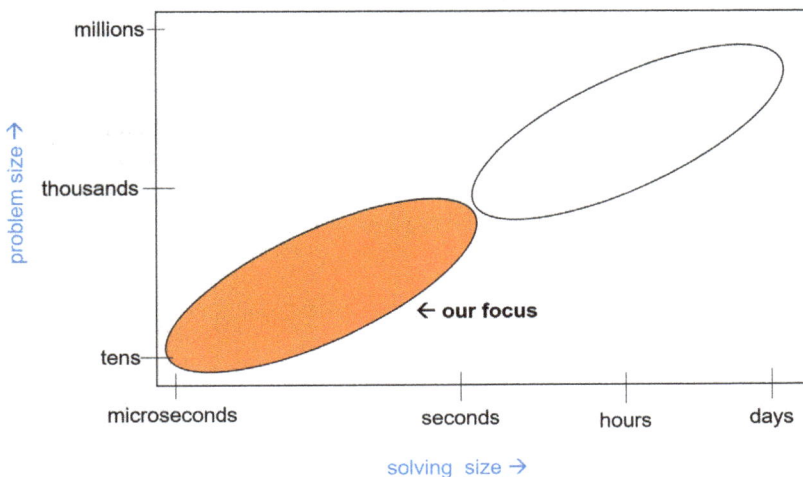

Figure 9.2   A view of optimisation problem size that we can focus on

or YALMIP. In fact, CVXPY solves problems in simple ways and parallelism constructs and can be extended for nonconvex optimisation in future. Similar to CVX, CVXPY is also easy to learn and use via many examples available.

9.2.3   Embedded in R programming

In R package, CVXR [224] is an integrated tool providing the object-oriented modelling language for convex optimisation, similar to CVX and CVXPY. Without restrictions to a standard form of solvers, CVXR users can formulate convex optimisation problems in a natural mathematical model by combining constants, variables and parameters using a mathematical properties library. Once having verified the problem's convexity by disciplined convex programming, CVXR transforms the problem into a standard conic form and passes it to one of the solvers such as ECOS or SCS solvers.

9.2.4   Embedded in JULIA programming

Solving optimisation in the high-level Julia programming is called 'JuliaOPT'. It includes Convex.jl [225] and JuMP.jl [226] packages.

Convex.jl (CVX.jl) is a Julia package for solving linear programs, mixed-integer linear programs and DCP-compliant convex programs under disciplined convex programming, through the MathProgBase interface. Convex.jl transforms the problem (objective functions and constraints) into a cone program and passes the problem to one of the solvers. The following functions are to form new convex, concave or affine expressions. Depending on the type of variables and problems, the conic constraints belong to linear, second-order, exponential, semidefinite constraints or integer (or binary) constraints.

JuMP is a domain-specific Julia modelling language for embedded optimisation in mathematics. JuMP supports a number of open-source solvers for a variety of problem classes, e.g., linear programming, (mixed) integer programming, second-order conic programming, SDP and nonlinear programming with many popular optimisation solvers including Cbc, Clp, ECOS, GLPK, Gurobi, MOSEK, NLopt and SCS. JuMP can be easily embedded in complex workflows including simulations and web servers. In fact, JuMP specifies and solves optimisation problems in simple ways that do not require expert knowledge. However, for advanced development, it can also allow experts to implement advanced algorithmic techniques such as exploiting efficient methods in linear programming or novel branch-and-bound solvers.

Chapter 10

Real-time embedded optimisation in UAV communications

Unmanned aerial vehicles (UAVs) have been emerging to become a major trend in the next generation of wireless networks [227–231]. With flexible configuration and mobility, UAVs can be more efficient and inexpensive for deployment, which is critical in Internet-of-Thing (IoT) applications. UAVs can easily gather information, manipulate physical objects or engage some equipment in remote or dangerous places. With a wide variety of vehicle types, UAVs can operate not only on the ground but also in a variety of environments such as space, air, water or underground. As a result, there are many types of applications where UAVs can be exploited such as environmental remediation, navigation in order to gather data, military applications, transportation of goods and performing dangerous tasks [232]. For instance, sensor nodes can be deployed with UAVs to estimate the path and velocity of tracked vehicles. For real-time applications, the tracked results will be collected by the UAVs and reported to the central system within strict time deadlines [233].

10.1   Unmanned aerial vehicle networks in IoT

10.1.1   Introduction to UAV in IoT

There are many issues needed solving in order to provide effective, stable and reliable UAV networks. The efficient deployment and operation in UAV communications is very important yet challenging to fulfil the promise of high capability and capacity [228]. The control of UAVs depends heavily on the sensors, communication devices, onboard processing power and existing appropriate models. Depending on terrain types (water, air or ground), a UAV must have its own modelling and operating system. The modelling of UAVs presents many challenges and has led to interest in areas such as nonlinear systems and artificial intelligence [232].

Airborne UAVs often move in organised swarms in three dimensions with rapid change in position. Hence, their topology could change in UAV networks. For example, the relative positions of UAVs may suddenly change, leading to some UAVs losing their trajectory and power, and, thus, they need to be brought down for redesigning and recharging. Furthermore, UAVs may malfunction and interrupt the network, prompting the network links to vanish, changing the positions of the nodes.

Thus, the design and deployment of trajectory optimisation is one necessary mission in UAV communication systems [234–238].

On the other hand, UAV devices typically have limited energy storage. The energy constraints will be much more important in small UAV networks, especially airborne UAVs. These UAVs may typically have only enough power for a flight with a few minutes or maximum an hour. Most signals are transmitted under lower power and thus the outage probability is high because the links might be intermittent. More importantly, these nodes would dynamically work with frequent reorganisation of the network, meaning that their routing protocols need to vary over time and they will use more energy to prolong the stability of the network. Therefore, the deployment and resource allocation such as spectrum or transmit power allocation should be considered in UAV-based applications.

A fully autonomous multi-UAV network will require robust inter-UAV cooperation in keeping the network organised. The ability to self-organise is essential for UAV networks to change the Media Access Control (MAC) and network layers, be tolerant to delays, flexible and automatic control through SDNs as well as employing efficient energy-saving schemes. Therefore, UAV systems must be more dynamic and intelligent. With efficient robust algorithms, low-complexity estimates and support for self-organising systems, machine learning will bring many benefits to UAV wireless networks.

Having discussed the above, it is obvious that optimisation, which is applied in the deployment of UAV communication, is indispensable. Moreover, real-time optimisation for embedded UAV systems is a promising research direction for modern wireless communications.

10.1.2   Characteristics of UAV networks

In this section, we consider the general characteristics of UAV-based wireless communication. The general communication model consists of cellular base stations, UAVs and ground terminals (GTs).

3D location

Within the completion time T and trajectory design, a UAV's location is denoted as $q(t) = [x(t), y(t), H(t)]^T \in \mathbb{R}^3$ at time t. We assume the initial and final time of the UAV's location are $t^{(0)} = 0$ and $t^{(N)} = T$. By introducing the elemental time-slot length δ_t, the horizon time T is divided into M time slots, e.g., $T = M\delta_t$. The elemental time-slot length is chosen so that the location of UAVs and GTs is approximately assumed as constants within each slot. Otherwise, the 3D location of the UAV can be rewritten as

$$q[m] = [x[m], y[m], H[m]]^T, \ m = 1, ..., M \tag{10.1}$$

Channel model

Due to the higher chance of LoS propagation via 3D location, the air-to-ground (ATG) channel differs from the terrestrial channel [239]. The effect of the environment on LoS occurrence becomes critical. The effects of propagation blockage [240], e.g., building blockage, still exist for the complete channel models. As a result, Rician, large-scale Rayleigh and free-space fading models are appropriate for ATG channels.

For an arbitrary elemental time slot, the distance between the UAV located at (x, y, H) and the kth GT is denoted as $R_k = \sqrt{d_k^2 + H^2}$, where $d_k = \sqrt{(x - x_k)^2 + (y - y_k)^2}$ is the horizontal distance between the UAV and GT k, and (x_k, y_k) is the 2D location of GT k. Furthermore, the altitude of the GTs and the antenna gains of both UAV and GTs are neglected. The distance path loss between the UAV and GT k is represented as [241]

$$L(R_k) = 10 \log \left(\frac{4\pi f_c R_k}{c} \right)^\alpha \tag{10.2}$$

where f_c is the carrier frequency (Hz), c is the speed of light (m/s) and $\alpha \geq 2$ is the path loss exponent.

On the other hand, the probability of LoS is given by [239, 242]

$$P_{LoS} = \frac{1}{1 + a \exp(-b(\arc \tan(\frac{H}{d_k}) - a))} \tag{10.3}$$

where a and b are constants depending on the environment. Thus, one has $P_{NLoS} = 1 - P_{LoS}$.

Then the total path loss expression from the UAV to the kth GT is

$$PL(R_k) \quad = \quad L(R_k) + \eta_{LoS}P_{LoS} + \eta_{NLoS}P_{NLoS} \quad = \quad 10 \log \left(\frac{4\pi f_c R_k}{c} \right)^\alpha + (\eta_{LoS} - $$

$$\eta_{NLoS}) \frac{1}{1 + a \exp(-b(\arc \tan(\frac{H}{d_k}) - a))} + \eta_{NLoS} \tag{10.4}$$

where η_{LoS} and η_{NLoS} are average additional losses for LoS and NLoS, respectively.

Free-space model

We assume LoS-dominated environment in free-space path loss model ($\alpha = 2$) [243]. Within an arbitrary time slot [t], the channel gain in the free-space model is defined as

$$g_k(t) = \beta_0 R_k^{-2}(t) \tag{10.5}$$

where β_0 is the channel power gain at the reference distance d_0,

$$R_k(t) = \sqrt{d_k^2(t) + H^2(t)}$$

is the distance between the UAV and kth GT at the time slot t.

For a constant transmit power (P_U) at the UAV, the received power at the GT k can be addressed as

$$P_k(t) = P_U g_k(t) \triangleq \frac{\beta_0 P_U}{(x(t) - x_k(t))^2 + (y(t) - y_k(t))^2 + H^2(t)} \tag{10.6}$$

Thus, the total received power at the GT k in the period of time (T) is given as

$$E_k = \int_0^T P_k(t)dt \tag{10.7}$$

Routing and trajectory design

We consider three fundamental aspects of UAV's trajectory including trajectory location $\{q[m]\}_{m=1}^M$, speed $\{v[m]\}_{m=1}^M$ and acceleration $\{a_{cc}[m]\}_{m=1}^M$.

With the UAV's fixed altitude at H, its trajectory location in the m time slot is simply the horizontal location $q[m] = [x[m], y[m]]^T$, $m = 1, ..., M$, where M is the final time slot at the end of trajectory.

The trajectory of UAV is defined as

$$q[m + 1] = q[m] + v[m]\delta_t + \frac{1}{2} a_{cc}[m]\delta_t^2, \ m = 0, 1, ..., M. \tag{10.8}$$

$$v[m + 1] = v[m] + a_{cc}[m]\delta_t, \ m = 0, 1, ..., M. \tag{10.9}$$

For a constant of velocity $(a_{cc}[m] = 0)$ and allowable maximum velocity of the UAV (V_{max}), the trajectory constraints of the UAV are

$$\|q[m + 1] - q[m]\| \leq V_{max}\delta_t, \ m = 0, 1, ..., M \tag{10.10}$$

where $q[0] = [x[0], y[0]]^T$ is the initial horizontal location of the UAV.

On the other hand, trajectory tracking is important in UAV applications. The general optimisation of trajectory tracking is to minimise the Euclidean distance between the UAV and the targets. The targets can be either UAV devices or GTs.

$$\min_{x,y,z} \ (J_x, J_y, J_z) \tag{10.11a}$$

$$\text{s.t.} \quad \|q[m + 1] - q[m]\| \leq v[m + 1]\delta_t, \ m = 0, 1, ..., M - 1, \tag{10.11b}$$

$$v[m] \leq V_{max}, \ m = 1, ..., M. \tag{10.11c}$$

where $J_x = \sum_{i=1}^N (x - x_i)^2$, $J_y = \sum_{i=1}^N (y - y_i)^2$ and $J_z = \sum_{i=1}^N (H - h_i)^2$. (x_i, y_i, h_i) denote the location of the ith target.

Model predictive control

In trajectory design, model predictive control (MPC) is tremendously required [244–246].

For a finite-time control $T = \sum_{m=0}^{M-1} m\delta_t = M\delta_t$, a general problem of MPC is given as

$$\min_{q,u} \ J(q, u) \tag{10.12a}$$

s.t. $\quad q_{m+1} = Aq_m + Bu_m, \quad m = 0, 1, ..., M - 1,$ (10.12b)

$\quad\quad q_m = q[m], u_m = u[m], \quad m = 0, 1, ..., M - 1.$ (10.12c)

where $J(.)$ is the cost function. A and B are information matrices with respect to trajectory design q and decision control variables u, respectively. An example of cost function $J(.)$ can be expressed as

$$J(q, u) = \|x_M\|_{P_M}^2 + \sum_{m=0}^{M-1} \|x_m\|_Q^2 + \|u\|_R^2 = x_M^T P_M x_M + \sum_{m=0}^{M-1} x_m^T Q x_m + u^T R u \quad (10.13)$$

where P_M, Q and R are the final-state cost, stage-state cost and input-state cost of the MPC problem, respectively. The control input at time slot m is denoted as $u_m = \{u[m], u[m + 1], ..., u[M - 1]\}$, where $u[m]$ satisfies the lower bound and upper bound of control input as $u[m] \in [\underline{u}, \overline{u}], \quad m = 0, ..., M - 1$.

An example of a MPC problem in the trajectory tracking application is given as

$$\min_{q, u} \quad J(q, u) = J_1(q) + J_2(q) + J_3(u)$$ (10.14a)

s.t. $\quad q \in \mathcal{Q}$ (10.14b)

$\quad\quad \max(u_{min} - \overline{u}, -\sigma_u) \leq u \leq \min(u_{max} - \overline{u}, \sigma_u).$ (10.14c)

where $\mathcal{Q}$ is a convex set of trajectory design. u_{max}, u_{min} and $\overline{u}$ are the upper bound, lower bound and average of control input. σ_u is the error control. $J_1(q)$, $J_2(q)$ and $J_3(u)$ represent the path-tracking error, the square of error and the excessive control and oscillation.

10.1.3 Design and management of UAV systems

Figure 10.1 illustrates an embedded convex optimisation system in UAV-enabled communication. A practical optimisation problem is an input to the system. Efficient methods are used for solving the problem on computer ((central processing unit (CPU), graphics processing unit (GPU), Field Programmable Gate Array (FPGA)), applying the novel approaches for real-time optimisation as provided in Chapter 5, e.g., first-order methods, parallelism approaches and learning-based optimisation algorithms. A custom code is produced by automatic code generation by an embedded software before it is embedded into the real system. On the other hand, convex optimisation problems can be solved manually by transforming the problems into standard optimisation programming. In fact, these problems are often small and simple. The manual solution will take more tasks when new constraints or objectives are added to the problem. Meanwhile, the embedded optimisation programming, e.g., CVX, CVXPY and Julia OPT, does it automatically. Hence, large-scale and complicated problems will be much clearer in embedded programming.

For an example of embedded UAV communication, a Raspberry module can be integrated into an UAV kit for controlling UAV trajectory and performing embedded optimisation module. Figure 10.2 presents a general model of UAV communication where multiple embedded UAV kits operate and exchange via backhaul connections of the network information, e.g., channel state information, energy harvesting and auto-code generation of embedded convex optimisation. The ground station can be

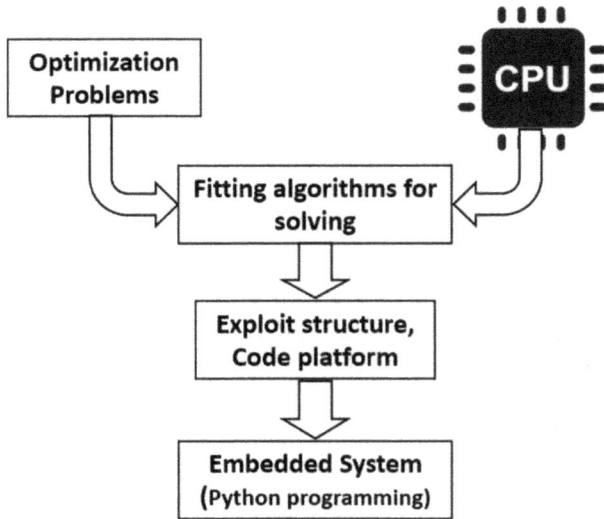

Figure 10.1   Embedded optimisation system in UAV-enabled communication

represented as fog computing for controlling a group of UAVs in IoT applications. One such example is the use of machine learning in UAV networks, a system based on learning optimisation algorithms that now forms a key technological platform for real-time scenarios. In Figure 10.3, the UAV can apply real-time optimisation

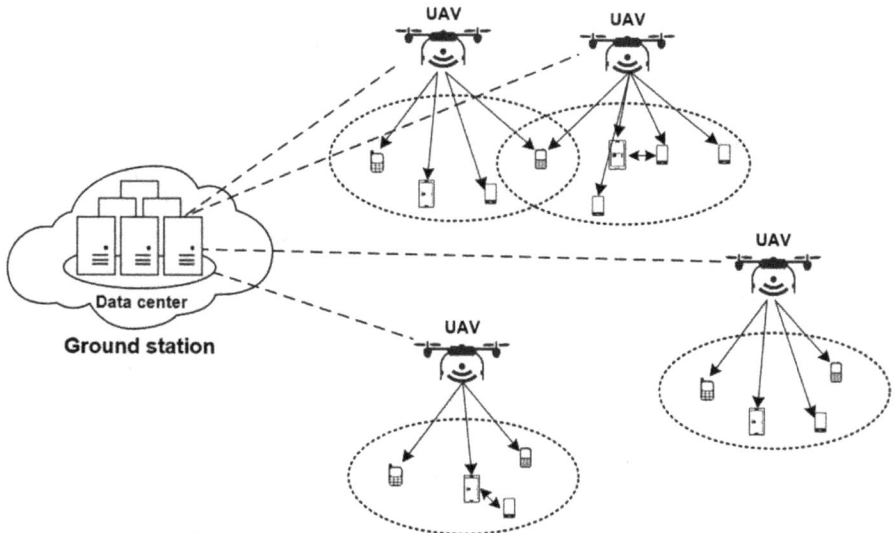

Figure 10.2   UAV communication for IoT application with fog computing as a ground station and multiple embedded drone stations

Ground station
Powerful computer
GPU, CPU, FPGA ...

Wireless channel

Flying station
UAV kits
Embedded module control
(Arduino, RaspPi, Arm ...)

Training stage
➤ Offline, online training
➤ Learning-based OPT Alg.
➤ Trained neural net

Testing stage
➤ RT-OPT
➤ Applications

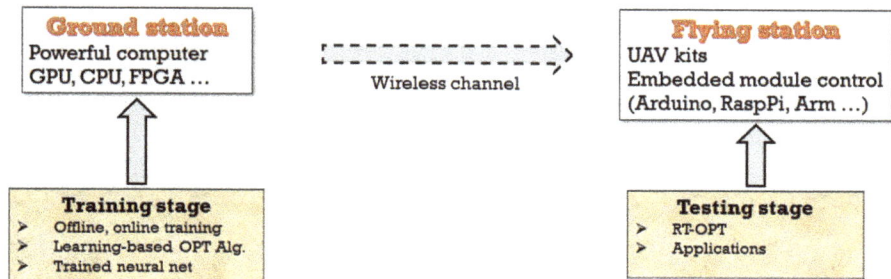

Figure 10.3   *Training and testing stages for applying machine learning in UAV communication*

applications at the testing stage with learning-based optimisation algorithms applied at the training stage in the ground station. There are many open issues for applying machine learning and embedded optimisation algorithms in wireless communication networks. Several important key technologies for this trend are introduced in Chapter 8.

We now present an efficient approach to implement the real-time optimisation applications via the collaboration between the ground station and flying stations (UAV kits). As shown in Figures 10.3 and 10.4, an embedded real-time optimisation package will be performed at the ground station with novel approaches such as first-order approximation methods discussed in Chapter 6, distributed and parallel computing discussed in Chapter 7 and machine learning discussed in Chapter 8.

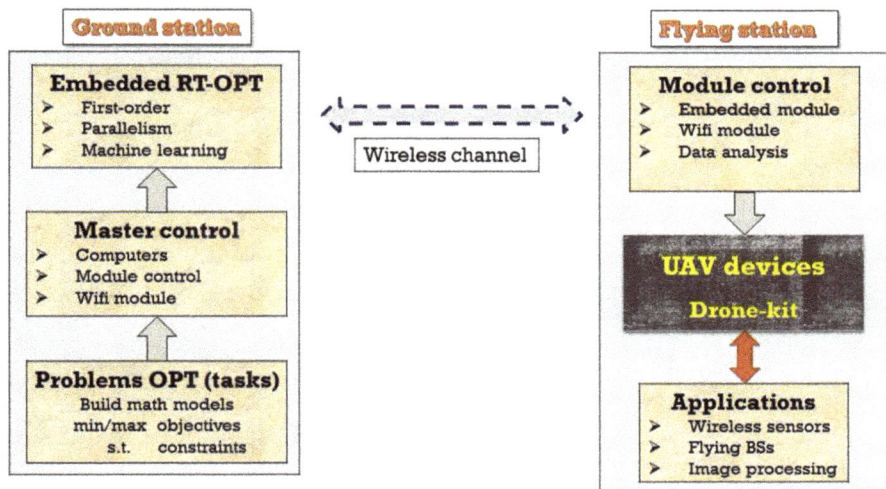

Ground station

Embedded RT-OPT
➤ First-order
➤ Parallelism
➤ Machine learning

Wireless channel

Flying station

Module control
➤ Embedded module
➤ Wifi module
➤ Data analysis

Master control
➤ Computers
➤ Module control
➤ Wifi module

UAV devices
Drone-kit

Problems OPT (tasks)
Build math models
min/max objectives
s.t.   constraints

Applications
➤ Wireless sensors
➤ Flying BSs
➤ Image processing

Figure 10.4   *A model of real-time UAV communication system with the ground station and the flying station.*

Chapter 11

An introduction of real-time embedded optimisation programming for UAV systems

For disaster communications,[a] it is very challenging for the contemporary wireless technology and infrastructure to meet the demands for connectivity. Modern wireless networks should be developed to satisfy the increasing demand for quality-of-service (QoS) in mission-critical communications for disaster management, which are currently faced with the challenges of limited spectrum, expensive resources, reliable and green communication. There is a tremendous need for optimisation techniques in the study and design of the key functionalities of wireless systems. Until now, almost all current optimisations are often carried out on large timescales (e.g., minutes or hours) without strict time constraints for solving the problems. With the improvement of computational speed, efficient algorithms and advanced coding approaches, a framework of real-time optimisation programming, which plays a major role in the trend of modern engineering such as mission-critical communications, is introduced in the context of natural disaster. In particular, this chapter gives an introduction of embedded convex optimisation programming for unmanned ariel vehicle (UAV) communications in disaster networks with strict supervision on execution time in real-time scenarios.

11.1   Introduction

It is often a formidable task to promptly restore and sustain communication networks that are damaged during a disaster [247, 248]. In addition, the entire area will need to be accurately evaluated in real-time, which is challenging as well [249, 250].

   To support communication and situation evaluation in disaster management, emerging wireless technologies [251–255] based on Internet-of-Things (IoT) platform consisting of device-to-device (D2D) communications, wireless sensor networks (WSNs) and UAV-enabled communication [256, 257] controlled by ground

---

[a]This chapter has been published as part of [97].

control stations (GCSs) are used. With flexible configuration and mobility, UAVs are very efficient and inexpensive for deployment in disaster-prone areas [258, 259]. UAVs can easily gather information, manipulate physical objects, or engage equipment in remote or dangerous places [260–262]. In order to provide adequate network coverage for particular areas affected by the disaster, there are many optimisation problems that should be considered and solved in real-time scenarios for exploiting and controlling the UAV systems. This will ensure ultra-low network latency in transmission traffic and enable real-time applications in disaster scenarios [250, 259]. These optimisation problems have often been associated with many efficient (linear and nonlinear) methods and algorithms such as filter design, resource allocation and transmission strategies. These methods are chosen such that the network performance is optimised corresponding to the demand of network services via optimisation problems.

In many scenarios of communication system with a constantly changing environment, optimisation applications are used to optimise the problems that need to be solved in online applications, i.e., channel estimation, dynamic resource allocation and video streaming. The strict real-time deadline is the most important requirement to be met in online applications. In all current optimisation scenarios, the processing time for solving convex optimisation problem is often in minutes or hours, or even longer. This is because traditional convex optimisation methods are computationally expensive, requiring longer solving time. Further to this, modern optimisation problems are potentially formed as expensive and complex problems such as the combination of continuous and integer decision variables, multi-objective, large-scale problems and unknown objective information.

Fortunately, the rapid increase of powerful computers, efficient algorithms and novel approaches has promised an enormous decrease in calculation time in optimisation algorithms for embedded convex optimisation programming, which can solve modern optimisation problems with strict deadlines of micro-seconds or milliseconds [74, 259]. In fact, some modest-sized optimisation problems can be handled by online algorithms at fast processing time and within a given sufficient amount of computer memory [74, 259]. Moreover, in the near future, efficient methods using parallel computing are applicable to real-time applications for solving large-scale problems. Another potential approach for real-time optimisation is to interplay the frontier of modern optimisation and machine learning techniques such as low-rank minimisation, sparsity methods, decomposition techniques or statistically learning models.

11.2 UAV-enabled communication networks

UAVs have been emerging as a major trend in the next-generation wireless networks [228–230]. With a wide variety of vehicle types and operational environments, UAVs can be easily deployed on the ground, in the air, water or underground

[232]. As a result, there are many types of applications that UAVs can be exploited such as environmental remediation, navigation to gather data, military applications, goods transportation, performing dangerous tasks and so on [232]. In addition, UAVs can be utilised for governmental or non-governmental purposes in rescue operations such as public protection and disaster recovery. Some good examples are surveillance and reconnaissance, homeland security, environmental monitoring, agriculture and construction industry. For a real-time application, tracking results will be collected at the UAVs and reported to the central system within strict time deadlines [233].

In natural disasters, keeping communication connectivity provides a lifeline. The lack of communications in remote areas and poor conditions of communications in developing countries can have the detrimental effects. As one of the most feasible technologies, online algorithms could be integrated into UAVs. This could be of great help when networks are congested, and there is a lack of power supply. Therefore, real-time optimisation in UAV communications to tackle time constraint will play a crucial role in disaster scenarios. In these cases, UAVs that are flying above the affected area could help first responders to assess the situation and deliver a coverage-based cellular network as quickly as possible. To develop practical optimisation in emergency situations, developers should focus on reducing the execution time and computational complexity of optimisation programming in real-time scenarios.

11.2.1 Challenges of UAV-enabled communications

There are many issues to solve in order to provide effective, stable and reliable UAV networks. The efficient deployment and operation of UAV communications are very important, yet challenging, to meet the promising high capability and capacity [228]. The control of UAVs might depend heavily on sensors, communication devices, onboard processing power and appropriate existing models. The modelling of UAVs has led to an interest in areas such as non-linear systems and artificial intelligence [232]. Airborne UAVs often move in organised swarms in three dimensions with a rapid change of network trajectory. For example, a change in the relative positions of UAVs may lead to some UAVs losing their trajectory and power, and, thus, they need to be brought down for redesigning and recharging. Furthermore, UAVs may malfunction or interrupt the network, causing the network links to vanish and changing nodes' positions [234, 236, 237].

On the other hand, UAV devices typically have limited energy storage for flying operations. The energy constraints will be much tighter in the long-term operation of UAV networks. Most signals are transmitted under lower power and thus the outage probability is high because the links might be intermittent. More importantly, these nodes would dynamically work to frequently reorganise the network. This means that their routing protocols need to verify changes over time and use more energy to

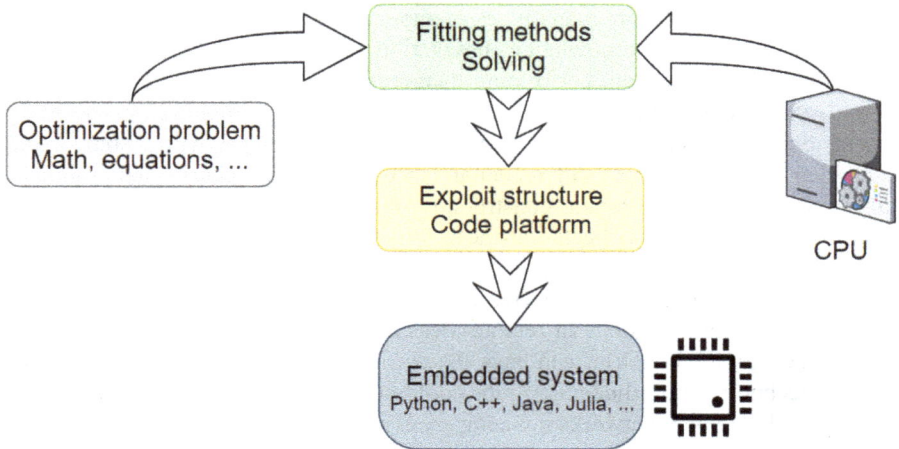

*Figure 11.1   An embedded optimisation system in UAV-enabled communication.
©EAI 2018. Reprinted with permission from [97]*

prolong the network stability. Therefore, deployment and resource allocation should be considered in UAV-based applications.

A fully autonomous multi-UAV network will require a robust inter-UAV network to cooperate in keeping the network organised. The ability of self-organising is essential in UAV networks to change the operation models, network and application layers, be tolerant to delays to enable flexible and automatic control as well as employ efficient energy-saving schemes. Therefore, the UAV systems must be more dynamic and wise. Machine learning is a potential trend for UAV networks by lending efficient robust algorithms, low-complexity estimates and supporting self-organising systems. Finally, real-time optimisation for embedded UAV systems is a promising research direction for modern wireless communications.

11.2.2   Practical embedded optimisation programming for UAV systems

In [74], embedded convex optimisation programming is able to solve many optimisation problems at fast timescales. In fact, in real-time scenarios, solving optimisation problems does not only have to yield an optimal feasible point but also meet a strict time limit (deadline time constraint). A diagram of embedded convex optimisation system is given in Figure 11.1. Efficient methods are used for solving the problem which is exploited on a type of computer (central processing unit (CPU), graphics processing unit (GPU), Field Programmable Gate Array (FPGA)) applying the novel approaches for real-time optimisation

Figure 11.2   *A practical model of UAV communication system with a ground station and a flying station. ©EAI 2018. Reprinted with permission from [97].*

programming. Finally, a custom code platform is produced by automatic code generation by an embedded software.

For instance, Figure 11.2 presents a general model of UAV communication where multiple embedded UAV kits operate and exchange the network information via backhaul connections, e.g., channel state information, energy harvesting and auto-code generation of embedded convex optimisation. The ground station can be represented as central computing for controlling UAV communication network. The UAVs can apply real-time optimisation applications at the testing stage with learning-based optimisation algorithms applied at the training stage in the ground station.

11.3   Practical applications for embedded optimisation in UAV systems

To give the reader a better appreciation of the performance of embedded optimisation, the performance evaluation of a UAV network is implemented in Python and MATLAB® language programming. For embedded optimisation applications, simulation results are implemented by using CVXPY 1.0.6 package [223] in Python and CVX 2.1 [221] in MATLAB. We also provide two computational platforms (PL) for performing optimisation operations, as follows:

- PL1: A laptop with Intel Core(TM) i7, CPU @2.80 GHz and 16 GB memory
- PL2: A Raspberry Pi 3B module with a 1.2 GHz Quad-core 64-bit CPU ARM cortex-A53 and 1 GB RAM.

EXAMPLE 1 (Optimal network coverage)

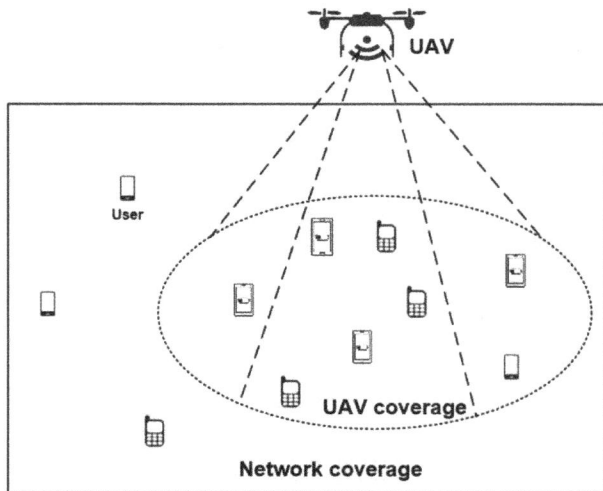

Figure 11.3   Network coverage of a UAV. ©EAI 2018. Reprinted with permission
            from [97].

In the first example, the location of the UAV should be optimised for supporting
as many ground terminals (GTs) as possible. The considered model is provided in
Figure 11.3. The system consists of an UAV with the fixed altitude H and K users
(UEs), which are randomly distributed in a given network coverage.

First, we assume the 3D location of the UAV is (x, y, H). The location of a GT
is as (x_k, y_k) for $k = 1, ..., K$. We also define the ratio of altitude and network coverage of the UAV as

$$\phi = H/r_{cov} \tag{11.1}$$

where r_{cov} is the radius of the UAV's network coverage. In this work, we choose the
value of ϕ at 1 (e.g. unbar scenarios), which means $H = r_{cov}$.

Hence, the kth GT will be served by the UAV if

$$d_k^2 = (x - x_k)^2 + (y - y_k)^2 \leq r_{cov}^2 = H^2 \tag{11.2}$$

Our target is to maximise the number of GTs that the UAV can support at the same
time. The considered problem is given by

$$\max_{x,y,u_k} \sum_{k=1}^{K} u_k \tag{11.3a}$$

$$\text{s.t. } d_k^2 \leq r_{cov}^2 + \lambda(1 - u_k), \ k = 1, ..., K \tag{11.3b}$$

$$x_{min} \leq x \leq x_{max} \tag{11.3c}$$

$$y_{min} \leq y \leq y_{max} \tag{11.3d}$$
$$u_k \in \{0, 1\}, \quad k = 1, ..., K \tag{11.3e}$$

where λ is chosen as a specific value such that it is larger than the maximum network area that we consider. (x_{min}, x_{max}), (y_{min}, y_{max}) and (H_{min}, H_{max}) are the lower and upper bounds of the horizontal, vertical and altitude range of UAV, respectively. Note that the problem in (11.9) is mixed-integer (binary) quadratic programming which is nonconvex. Fortunately, Python-embedded optimisation programming (CVXPY) can solve the problem (11.9) effectively.

For simulation, we assume the coverage is a square area with the length of each side is 300 m. (x_{min}, x_{max}), (y_{min}, y_{max}) and (H_{min}, H_{max}) are set as $(-300, 300)$, $(-300, 300)$ and $(47, 122)$ in metres, respectively. There are $K = \{10, 20, 30\}$ UEs in the considered network. For this example, the problem (11.3) is transformed into a mixed-integer programming with 10, 20 and 30 binary variables and 15, 25 and 35 inequality constraints, respectively. Then, the Python code generated in PL1 by CVXPY solves this problem in approximately 69, 362 and 703 ms, respectively.

EXAMPLE 2 (Tracking and routing plan)

With mobility and flexible configuration, UAVs are developed for tracking targets or routing plan. In this example, the trajectory of a UAV is designed for data collection in a wireless sensor network. The considered model is provided in Figure 11.4, where a number of sensors are distributed in the network area.

We assume the UAV operates at the constant altitude H. There are total $K + 1$ time slots corresponding to K sensor nodes that have data to be collected and an initial time slot. At the initial point ($k = 0$), we assume the horizontal location of the UAV is $q[0] = [0, 0]$. Then, its location at the kth time slot is given as

$$q[k] = [x[k], y[k]]^T \tag{11.4}$$

The Euclidean distance between the UAV and sensor k at time slot k is

$$d_k[k] = \sqrt{(x[k] - x_k)^2 + (y[k] - y_k)^2} \tag{11.5}$$

where (x_k, y_k) represents the location of kth sensor node.

Each sensor should forward its data with size (D_k) at transmission rate (C_k) to the UAV. Thus, the UAV's flying time over sensor node k (T_k) for completing the data collection should follow the inequality

$$T_k \geq D_k/C_k , \quad k = 1, ..., K. \tag{11.6}$$

For convenience, the UAV's coverage is its altitude for urban scenarios. The UAV can collect the data from a sensor node only when the sensor node belongs to its coverage:

$$d_k[k] \leq r_{cov} = H , \quad k = 1, ..., K. \tag{11.7}$$

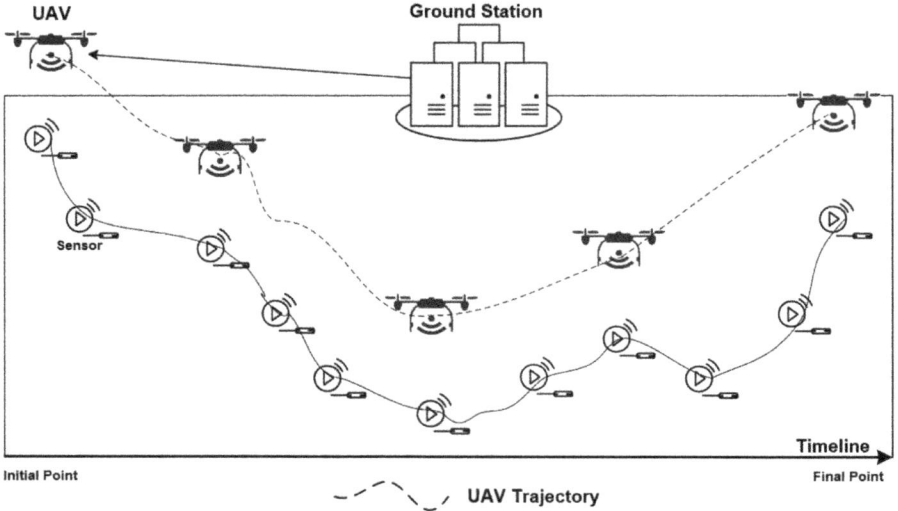

*Figure 11.4   A model of routing UAV for data collection from a sensor network.
©EAI 2018. Reprinted with permission from [97].*

Next, the following constraint of the UAV's flying time between two adjacent time slots is given as

$$\delta_k \geq \frac{\|q[k+1] - q[k]\|}{V}, \; k = 0, 1, ..., K-1 \tag{11.8}$$

where V is the constant flight speed of the UAV.

Our target is to minimise the total flying time that the UAV needs for collecting the data from K sensor nodes. The considered problem is given as

$$\min_{q, \delta} \sum_{n=1}^{K} \delta_k + T_k \tag{11.9a}$$

$$\text{s.t. } (11.6), (11.7), (11.8) \tag{11.9b}$$

$$x_{min} \leq x[k] \leq x_{max}, \; k = 1, ..., K \tag{11.9c}$$

$$y_{min} \leq y[k] \leq y_{max}, \; k = 1, ..., K \tag{11.9d}$$

For measurement simulation, the sensor nodes are horizontally distributed in the rectangle network area with the width and height of 5 000 m and 200 m, respectively. The altitude of the UAV is set as $H = 50$ m. The flight speed of the UAV is constant at all time with $V = 10$ m/s. The size data and transmission data rate at all the sensor nodes is the same at $D = 400$ bits and $C = 200$ bits/s. There are $K = [20, 60, 100]$ sensors in the considered network. For 100 sensor nodes, the problem (11.9) is transformed into a QP (quadratic programming)

Table 11.1   *Average execution time in different computational platforms for solving problem (11.8)*

# of UEs	Average flying time (s)	PL1 with CVXPY (ms)	PL2 with CVXPY (s)	PL1 with CVX (s)
20	530	138	1.25	2.19
60	625	397	3.41	4.04
100	740	638	5.86	6.2

with 300 variables and 700 inequality constraints. The Python code generated by CVXPY in PL1 solves the problem (11.9) in approximately 600 milliseconds while CVXPY in PL2 (Raspberry Pi module) and CVX in PL1 solve the problem in 6 seconds, as shown in Table 11.1.

EXAMPLE 3 (Discovery)

A UAV's trajectory can be designed for discovering many hard-to-reach places such as disaster-prone or extreme geographic areas. An example of the considered model is provided in Figure 11.5.

First, we separate the area into a set of sweeping rows. These rows represent the edges of a graph between two nodes at the edge of the area. The UAV with a mounted camera follows from the initial node to the end of a row and then turns around to the next row. In this direction, the area can be covered with the smallest number of rows, thus with the smallest number of curves. However, the sweep

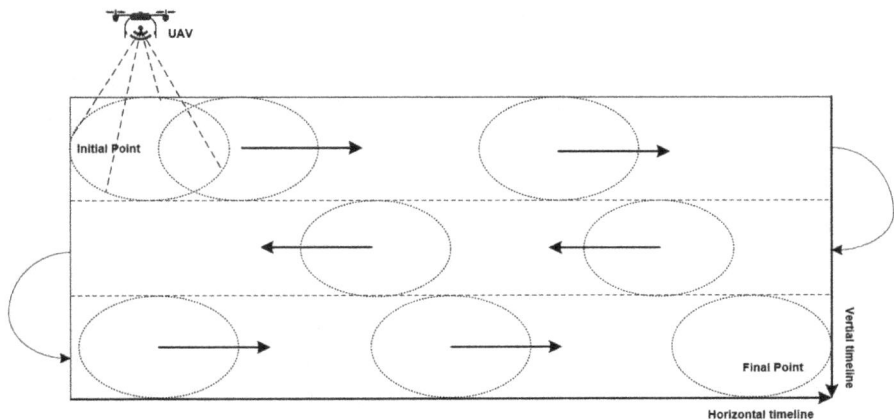

Figure 11.5   *A model of UAV trajectory for discovery missions. ©EAI 2018. Reprinted with permission from [97].*

direction can be chosen as a function of the environment (wind, weather) or characteristics of the considered area.

The altitude of the UAV is assumed constant and is chosen so that the camera resolution allows the best observation of the considered area on the ground. We assume that the camera has an image sensor with l_{cm} in width and a focal length of f_{cm}. The distance between the camera and the ground, H, is the altitude of the UAV. Thus, the footprint of the onboard camera on the ground L_{cm} is

$$L_{cm} = \frac{H l_{cm}}{f_{cm}} \tag{11.10}$$

Next, the distance between two adjacent rows is defined as a function of the footprint of the onboard camera on the ground. The number of coverage rows is given as

$$N_l = \left\lceil \frac{h_{min}}{L_{cm}(1 - s)} \right\rceil \tag{11.11}$$

whereas the distance between the two rows is

$$d_l = \frac{h_{min}}{N_l} \tag{11.12}$$

where $s \in (0, 1)$ represents the overlap fraction between two images.

We consider K UAVs for this discovering mission. The maximum flying time of the UAVs is finite and known in advance. Each UAV is equipped with an onboard camera. Before the flight, we consider the setup time of the UAVs as including the connection of components in the UAVs, fixing GPS, launching operation and other tasks. For presenting the mathematical formulation of the problem, the constant C_n defines the travel cost of row n. The binary variable $x_n^k \in \{0, 1\}$ is used to indicate whether or not the kth UAV flies in the nth row. V_n^k is the flight speed of the UAV k in row n. L_k is the battery duration of UAV k.

Thus, the time spent by the kth UAV to fly its route is

$$T_k = \sum_{n=1}^{N} \frac{C_n}{V_n^k} x_n^k + d_k \max \left\{ x^k \right\} \tag{11.13}$$

where d_k denotes the extra time to set up and launch UAV k, $x^k = [x_n^k]_{n=1}^{N}$. This means we do not need setup time for UAV k if the UAV is not used for discovering tasks, e.g., $\max\{x^k\} = 0$. We assume that all UAVs can depart at the same time. Our target is to discover the area on the ground using multiple UAVs. The deployment of UAVs is optimised for the discovering task so that the mission is completed in minimum time. Thus, the convex optimisation problem is as follows:

$$\min_{x} \sum_{k=1}^{K} T_k \tag{11.14a}$$

$$\text{s.t.} \sum_{n=1}^{N} \frac{C_n}{V_n^k} x_n^k \le L_k \ , \ k = 1, ..., K, \tag{11.14b}$$

$$\sum_{k=1}^{K} \sum_{n=1}^{N} x_n^k = N, \tag{11.14c}$$

$$\sum_{n=1}^{N} x_n^k \le 1 \ , \ k = 1, ..., K, \tag{11.14d}$$

$$x_n^k \in \{0, 1\} \ , \ n = 1, ..., N \tag{11.14e}$$

where the constraint (11.14b) is the maximum flying time constraint to guarantee the battery duration. To make sure the solution will have the UAVs cover the area modelled, the next constraint (11.14c) is also necessary. To guarantee that each node of the area is visited only once by a single UAV, the constraints (11.14d) are considered. Note that the problem in (11.14) is mixed-integer (binary) linear programming, which is a non-convex program. Fortunately, the problem can be effectively solved by the Python-embedded optimisation programming (CVXPY).

For simulation, we consider a rectangle searching area as (500, 4 000) m or 200 hectares, and the distance between two adjacent rows is $d_l = 100$ m. Thus, we have $N = 5$ rows to cover the considered area, e.g., $C_n = 4\,000$ m, $n = 1, ..., N$. $d_k = 5$ mins and $L_k = 20$ mins. The altitude of the UAVs is $H = 100$ m. With $K = 2$ UAVs ($K < N$), UAV 1 will serve the first three rows and UAV 2 will serve the remaining two rows, the solving time by Python-generated code with PL1 is approximately 25 ms. With $K = 6$ UAVs ($K > N$), the result is the same as the previous case, and, thus, the other UAVs, which are not used, can be reserved for other purposes. The solving time of the embedded optimisation in this case is approximately 57 ms.

EXAMPLE 4 (Relay-assisted networks)

With higher line-of-sight (LoS) probability, a UAV can be used as a relay node to serve multiple users who cannot be served by the base station (BS) due to block fading such as building blockage. Hence, there is no direct link between the BS and GTs. The location of the UAV is to be optimised for supporting multiple users. An example of the considered model is provided in Figure 11.6.

We also assume that the link between the BS and the UAV and the link between the UAV and the GTs are dominated by LoS links.

We assume the 3D location of the BS and the UAV is (x_{BS}, y_{BS}, H_{BS}) and (x_U, y_U, H_U), respectively. Then the distance between the BS and the UAV is expressed as

$$R_{BU} = \sqrt{d_{BU}^2 + (H_{BS} - H_U)^2} \tag{11.15}$$

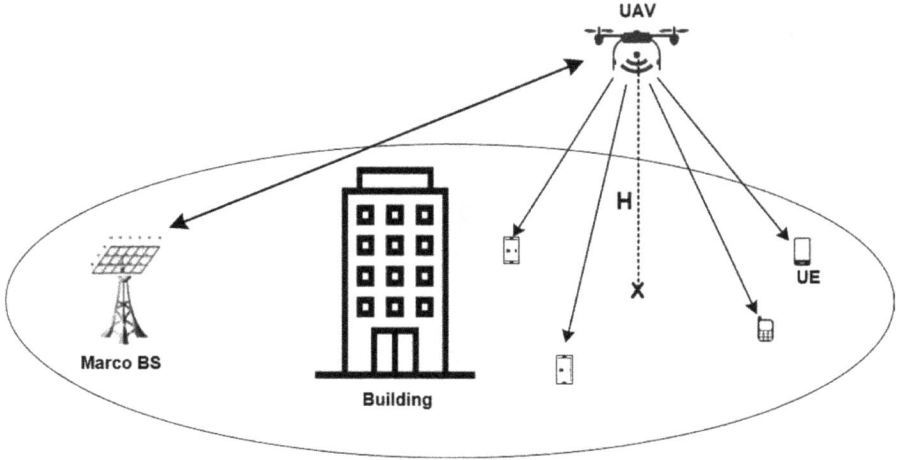

Figure 11.6   A relay network of an embedded UAV. ©EAI 2018. Reprinted with permission from [97].

where $d_{BU} = \sqrt{(x_U - x_{BS})^2 + (y_U - y_{BS})^2}$, H_{BS} and H_U are the antenna altitude of the BS and the UAV. We assume the antenna altitude of the UAV is also the altitude of the UAV. Thus, the free-space path loss channel can be used for the BS and the UAV link as follows:

$$g_{BU} = \beta_0 R_{BU}^{-2} = \frac{\beta_0}{(x_U - x_{BS})^2 + (y_U - y_{BS})^2 + (H_{BS} - H_U)^2} \tag{11.16}$$

And the free-space path loss channel for the UAV and kth GT link is given as

$$g_{Uk} = \beta_0 R_{Uk}^{-2} = \frac{\beta_0}{(x_U - x_k)^2 + (y_U - y_k)^2 + H_U^2} \tag{11.17}$$

where $R_{Uk} = \sqrt{d_{Uk}^2 + H_U^2}$, $d_{Uk} = \sqrt{(x_U - x_k)^2 + (y_U - y_k)^2}$ and (x_k, y_k) is the 2D location of the kth GT.

In this example, our target is to optimise the position of the UAV to maximise the total channel gain of the considered system including the BS-UAV link and UAV-GTs links. Hence, the considered problem is as follows:

$$\max_{x_U, y_U, H_U} \sum_{k=1}^{K} g_{Uk} + g_{BU} \tag{11.18a}$$

$$\text{s.t. } x_{min} \leq x_U \leq x_{max} \tag{11.18b}$$

$$y_{min} \leq y_U \leq y_{max} \tag{11.18c}$$

$$H_{min} \leq H_U \leq H_{max} \tag{11.18d}$$

However, problem (11.18) is nonconvex which is challenging to solve. By relaxing the objective function of problem (11.18), without loss of generality, problem (11.18) can be rewritten as the following convex problem:

$$\min_{x_U, y_U, H_U} \sum_{k=1}^{K} R_{Uk}^2 + R_{BU}^2 \tag{11.19a}$$

$$\text{s.t. } (11.18b), (11.18c), (11.18d) \tag{11.19b}$$

For simulation, we assume the coverage network with a circle radius of 500 m. The MBS is located at $(-300, 0, 30)$ while $K = \{10, 20, 30\}$ UEs are randomly distributed in the network coverage, but opposite to the MBS. (x_{min}, x_{max}), (y_{min}, y_{max}) and (H_{min}, H_{max}) are set as $(-500, 500)$, $(-500, 500)$ and $(47, 122)$ in m, respectively. For $K = \{10, 20, 30\}$ UEs, the problem (11.18), which is transformed into a QP programming, is solved by generating Python code in PL1 and PL2 in approximately $\{131, 325, 361\}$ ms and $\{1.15, 2.20, 3.33\}$ s using CVXPY and $\{1.54, 1.92, 2.35\}$ s using CVX in PL1, respectively.

EXAMPLE 5 (Cell-free networks)

With natural mobility and flexible configuration, UAVs can be effectively applied to cell-free networks [51]. In this example, we consider a cell-free massive MIMO network where K single-antenna GTs are served by $(47, 122)$ $(M \gg K)$ randomly deployed single-antenna UAVs as access points (APs) in the same time-frequency resource as illustrated in Figure 11.7. A central processing unit (ground station) connects to all the UAVs via a backhaul network for exchanging the network information, i.e. the channel estimates, precoding vectors and power control coefficients.

Suppose that g_{mk} is the channel between the mth UAV and the kth user. As in [50], we adopt the following channel model $g_{mk} = \sqrt{\beta_{mk}} h_{mk}$, where $h_{mk} \in \mathcal{CN}(0, 1)$ represents the large-scale fading while $h_{mk} \in \mathcal{CN}(0, 1)$ is the small-scaling fading. The channel matrix between all the UAVs and GTs is denoted by $G \in \mathbb{C}^{M \times K}$. Following the ATG channel model in (10.4), the large-scale fading under the LoS or NLoS links is given by

$$\beta_{mk} = 10 \log(\frac{4\pi f_c R_{mk}}{c})^{\alpha_g} + (\eta_{LoS} - \eta_{NLoS}) \frac{1}{1 + a \exp(-b(\arctan(\frac{H_m}{d_{mk}}) - a))} + \eta_{NLoS} \tag{11.20}$$

where H_m is the altitude of UAV m, $d_{mk} = \sqrt{(x_m - x_k)^2 + (y_m - y_k)^2}$ and $R_{mk} = \sqrt{d_{mk}^2 + H_m^2}$. (x_m, y_m, H_m) and $\hat{g}_{mk}$ are the 3D location of the mth UAV and the 2D location of the kth GT.

The transmission between the UAVs and the GTs is implemented through time-division duplex protocol. Focusing on the downlink performance, each coherence interval of length τ is divided into two phases: uplink training and

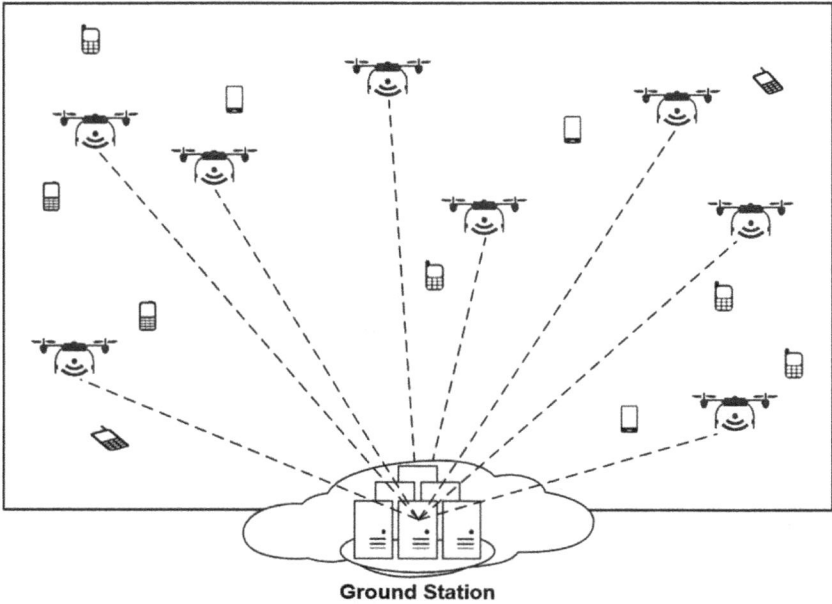

Figure 11.7   A model of UAV-enabled communication for cell-free network.
©EAI 2018. Reprinted with permission from [97].

downlink payload data transmission. In the first phase, on receiving the pilot signals from the GTs, each UAV estimates the channels to all the users via the minimum mean squared error technique. For perfect estimation, the channel estimation $\hat{g}_{mk}$ is defined as [51]

$$\hat{g}_{mk} \sim \mathcal{CN}\left(0, \frac{\rho_r \tau_u \beta_{mk}^2}{1 + \rho_r \tau_u \beta_{mk}}\right),$$ (11.21)

In the second phase, the APs use the channel estimates to precode and beamform data to all users. The received signal at the kth GT is given by

$$y_k = \sum_{m=1}^{M} g_{mk} x_m + n_k,$$ (11.22)

where $x_m = \sqrt{\rho_f} \sum_{k=1}^{K} \bar{f}_{mk} s_k$ denotes the transmitted signal of mth UAV to GTs, $x_m = \sqrt{\rho_f} \sum_{k=1}^{K} \bar{f}_{mk} s_k$ is the downlink power of each UAV, $E\{|s_k|^2\} = 1$ is the precoding coefficients and s_k is the symbol intended for the mth GT with $E\{|s_k|^2\} = 1$ and $n_k \in \mathcal{CN}(0, 1)$.

By applying zero-forcing processing for the transmission, $\bar{f}_{mk}$ can be expressed as

$$\bar{f}_{mk} = \sqrt{\eta k^b mk}, \, m = 1, ..., M, k = 1, ..., K, \tag{11.23}$$

where η_k is the power control coefficients and b_{mk} is the (m, k)th element of $\boldsymbol{B}$, where $\boldsymbol{B} = \hat{\boldsymbol{G}}^* (\hat{\boldsymbol{G}}^T \hat{\boldsymbol{G}}^*)^{-1} \in \mathbb{C}^{M \times K}$. Let $\bar{\boldsymbol{F}}$ is the precoding matrix whose $\bar{f}_{mk} = \sqrt{\eta_k} b_{mk}$, $m = 1, ..., M, k = 1, ..., K$,th element is $\bar{f}_{mk}$. The precoding matrix $\bar{\boldsymbol{F}}$ can be represented as $\bar{\boldsymbol{F}} = \boldsymbol{BP}$, where $\boldsymbol{P}$ is a diagonal matrix with $[\boldsymbol{P}]_{kk} = \sqrt{\eta_k}, k = 1, ..., K$.

The received signal at the kth user is

$$y_k = \sqrt{\rho_f} \hat{\boldsymbol{g}}_{[:,k]}^T \boldsymbol{BPs} + n_k = \sqrt{\rho_f} \sqrt{\eta_k} s_k + n_k. \tag{11.24}$$

Hence, the spectral efficiency of kth GT is

$$r_k = (1 - \frac{\tau_u}{\tau}) \log_2 \left(1 + \rho_f \eta_k \right) \tag{11.25}$$

where $\eta_{NLoS} = 23$ is the length of coherence interval slot for the uplink training which requires $\tau_u \geq K$ for perfect channel estimation.

For $\boldsymbol{\eta} = [\eta_k]_{k=1}^K$, the convex power minimisation problem is formulated as

$$\min_{\boldsymbol{\eta} > 0} \sum_{m=1}^{M} \sum_{k=1}^{K} \theta_{mk} \eta_k \tag{11.26a}$$

$$\text{s.t. } r_k \geq \bar{r}_k, k = 1, ..., K, \tag{11.26b}$$

$$\sum_{k=1}^{K} \theta_{mk} \eta_k \leq 1, m = 1, ..., M, \tag{11.26c}$$

where constraint (11.26b) represents the QoS requirement for each user and (11.26c) is the power constraint at each UAV, i.e., $E\{\|x_m\|^2\} \leq \rho_f$, where 1×1 is the ith element of $\boldsymbol{\theta}_{[m,:]}$ with

$$\boldsymbol{\theta}_{[m,:]} = \text{diag} \left\{ E \left((\hat{\boldsymbol{G}}^T \hat{\boldsymbol{G}}^*)^{-1} \hat{\boldsymbol{G}}_{[m,:]}^T \hat{\boldsymbol{G}}_{[m,:]}^* (\hat{\boldsymbol{G}}^T \hat{\boldsymbol{G}}^*)^{-1} \right) \right\}. \tag{11.27}$$

In the simulation, we consider an area of $1 \times 1 \, \text{km}^2$ with wrapped-around technique to avoid the boundary effects. All the UAVs and GTs are distributed randomly within the area. We choose the carrier frequency $\boldsymbol{\theta}_{[m,:]}$ GHz, bandwidth $B = 20$ MHz, $\tau = 200$ and ρ_f samples. The maximum transmit power of each AP (ρ_f) and user (ρ_r) are 200 mW and 100 mW. The noise power at the receivers is $N_0 = 290 \times \kappa \times B \times NF$, where κ and NF are Boltzmann constant and noise figure at 9 dB, respectively. ATG channel parameters are set as $a = 12.08$ and $b = 0.11$. The average additional losses for LoS and NLoS are set as $\eta_{LoS} = 1.6$ dB and $\eta_{NLoS} = 23$ dB.

For this example, the automatic code generation by CVXPY solves the afore-mentioned problem in milliseconds with the power minimisation in the considered

Table 11.2   The average solving time for the problem (11.24) in different computational platforms

# of UEs	PL1 with CVXPY (ms)	PL2 with CVXPY (s)	PL1 with CVX (s)
M=40, K=20	88	0.78	2.12
M=60, K=20	124	1.15	2.68
M=80, K=20	177	1.63	3.33
M=60, K=40	126	1.20	4.49
M=80, K=40	169	1.55	5.66
M=100, K=40	211	1.95	7.01

cell-free network, as shown in Table 11.2. Meanwhile, CVX tool takes a longer time to solve the same problem.

11.4   Conclusions

We have introduced the practical optimisation programming related to real-time applications and proposed the basic theory of the UAV systems based on the development of optimisation for disaster communications in real-time scenarios. This chapter opens a novel trend of optimisation for modern wireless communication systems. Based on this research, there are many issues and future works to study and necessarily call for novel solutions in the real-time optimisation for critical-mission wireless networks.

Chapter 12

Real-time optimal resource allocation for embedded UAV communication systems

This chapter considers device-to-device (D2D)[a] wireless information and power transfer systems using an unmanned aerial vehicle (UAV) as a relay node. As the energy capacity and flight time of UAVs are limited, a significant issue in deploying the UAV is to manage energy consumption in real-time application, which is proportional to the UAV's transmit power. To tackle this important issue, this chapter develops a real-time resource allocation algorithm for maximising the energy efficiency (EE) by jointly optimising the energy-harvesting time and power control for the considered D2D communication embedded with the UAV. This chapter demonstrates the effectiveness of the proposed algorithms as running time for solving them can be conducted in milliseconds.

12.1 Introduction

UAV-based communication networks with their flexible configuration and mobility are more efficient and inexpensive for the deployment of future wireless network [228] and the IoT applications [263]. Moreover, they are capable of enhancing wireless communications by virtue of the dominant presence of line-of-sight (LoS) connections [228]. Therefore, UAVs can provide novel schemes to enhance the network coverage for serving more wireless devices. A major issue, however, is that UAV devices typically have limited energy storage for flying operations, thereby the deployment and resource allocation such as spectrum or transmit power allocation should be considered for efficient utility [264, 265]. There are only a few existing works that have focused on the resource allocation aspect to improve the EE performance of UAV-based networks [266].

Although UAVs have been widely recognised as a promising technology to improve wireless networks performance their fundamental potential has not fully been exploited. An interesting development in UAV-based networks is the

---

[a]This chapter has been a part of [263].

integration of wireless power transfer (WPT). As a matter of fact, WPT in radio frequency has recently promised advanced technology for providing energy to wireless devices over the air (see e.g. [267] and the references therein). Very recently, WPT for UAV-enabled D2D networks has been considered in [268], where UAVs can operate as an energy supplier for multiple D2D pairs. Nevertheless, this work only considers throughput maximisation and does not focus on the aforementioned EE problem, which is crucial for providing efficient and lasting operation. To tackle this issue, we address the EE problem in the scenario of UAV-based relay networks supporting energy harvesting-enabled D2D communications. In particular, we investigate the issues of not only power allocation but also energy harvesting time, which will be formulated as a joint optimisation problem in energy harvesting-powered D2D communications. Nevertheless, joint optimisation problems are often complicated, for which we propose low-complexity efficient resource allocation algorithms.

Another critical issue in UAVs is the real-time control and operation due to their lifetime and dynamic environment. As such, we study the resource allocation problem of energy harvesting-powered D2D communications underlying UAV networks using real-time optimisation. The rapid improvement of computational speed as well as the use of efficient algorithms and advanced coding approaches in embedded convex approach allows optimal resource allocation problems to be solved in microseconds or milliseconds within strict time limits [74]. A lot of resource allocation optimisation problems have been considered in wireless communications. However, there is still a lack of investigation for real-time optimisation problems. With the exceeding development of computing performance, solving optimisation problems in real-time has become a requirement of wireless communication. This chapter considers real-time optimisation for resource allocation for embedded UAV-based communication systems.

12.2   Problem statement

Consider a communication system with one UAV serving multiple energy harvesting-powered D2D pairs as shown in Figure 12.1. The UAV and users are equipped with single antennas. Given a unitary communication time slot, the energy harvesting and information transmission in this UAV D2D network occur in two phases. In the first phase spanning τ with $0 < \tau < 1$, the dedicated D2D-transmitter (D2D-Tx) harvests energy from the UAV. Then, in the second phase spanning $(1 - \tau)$ the information transmission happens between D2D pairs.

The set of D2D pairs is denoted by $\mathcal{N} = \{1, 2, ..., N\}$. The energy harvested at the nth D2D-Tx is given by

$$E_n = \tau \eta P_0 g_n \tag{12.1}$$

where $0 < \eta < 1$ is the energy harvesting efficiency, P_0 is the maximum total transmit power at the UAV and g_n is the channel power gain from the UAV to the nth D2D-Tx.

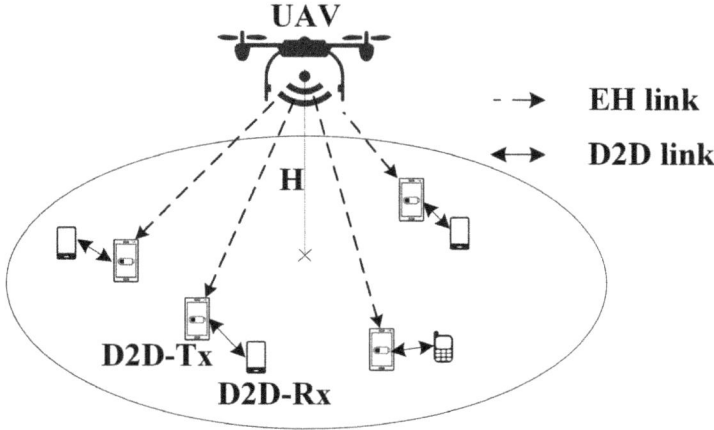

Figure 12.1   D2D communications assisted by a UAV. ©IEEE 2018. Reprinted with permission from [263].

For practical purpose, each user is assumed to utilise the harvested energy for information transmission phase. Denote by p_n the transmission power of the nth D2D pair. The following energy causality constraint must be satisfied

$$(1 - \tau)p_n \leq \tau \eta P_0 g_n, \ n \in \mathcal{N} \tag{12.2}$$

For $\mathbf{p} = [p_n]_{n=1}^N$, the information throughput (in nats) at the nth D2D pair is

$$r_n(\tau, \mathbf{p}) = (1 - \tau) \ln \left(1 + \frac{p_n h_{n,n}}{\sum_{i \neq n}^N p_i h_{n,i} + \sigma^2} \right) \tag{12.3}$$

where $h_{n,i}$ is the channel gain for the link from the nth D2D-Tx to ith D2D-receiver (D2D-Rx) and σ^2 is the noise power. Next, the total power consumption in the considered D2D network can be written as

$$\vartheta(\tau, \mathbf{p}) = \sum_{n=1}^N (1 - \tau)p_n + \tau \eta P_0 + P_{\text{cir}} \tag{12.4}$$

where P_{cir} is the circuit non-transmit power at the UAV.

In this chapter, the main target is to maximise the EE of the UAV network while satisfying the energy causality constraint and quality-of-service (QoS) constraint for each D2D pair. As such, the EE maximisation problem is as follows

$$\max_{\tau, \mathbf{p} > 0} \phi = \frac{\sum_{n=1}^N r_n(\tau, \mathbf{p})}{\vartheta(\tau, \mathbf{p})} \quad \text{s.t. (12.2)} \tag{12.5a}$$

$$r_n(\tau, \mathbf{p}) \geq \bar{r}, n \in \mathcal{N}, \tag{12.5b}$$

$$0 \leq \tau \leq 1, \tag{12.5c}$$

where the rate threshold $\bar{r}$ represents the QoS constraints.

Note that the problem in (12.5) is nonconvex because of the nonconcave objective functions (12.5a) and the nonlinear constraint (12.5b). In the next section, we propose a novel optimisation algorithm for solving the problem (12.5) in real-time.

12.3   Joint harvesting time and power allocation for EE maximisation

In this section, we propose a practical algorithm for the EE maximisation problem (12.5) by jointly optimising the energy harvesting time and power allocation. To solve the problem (12.5), we first change the variables [269]:

$$1 - \tau = \tfrac{1}{\theta} \text{ and } p_n \to \tfrac{1}{p_n}, \ n = 1, ..., N \tag{12.6}$$

such that the variable satisfies the convex constraint

$$\theta > 1. \tag{12.7}$$

Then, the problem (12.5) is equivalent to

$$\max_{\theta, \mathbf{p}} \phi = \frac{\sum_{n=1}^{N} r_n(\theta, \mathbf{p})}{\vartheta(\theta, \mathbf{p})} \text{ s.t. (12.7)} \tag{12.8a}$$

$$1/p_n \le (\theta - 1)\eta P_0 g_n, \ n \in \mathcal{N} \tag{12.8b}$$

$$\frac{1}{\theta} \ln\left(1 + \frac{h_{n,n}}{p_n \sum_{i \ne n}^{N} h_{n,i}/p_i + p_n \sigma^2}\right) \ge \bar{r}, \ n \in \mathcal{N}. \tag{12.8c}$$

where $\vartheta(\theta, \mathbf{p}) = \sum_{n=1}^{N} 1/(\theta p_n) + (1 - 1/\theta)\eta P_0 + P_{\text{cir}}$.

To solve the problem (12.8), we use the logarithmic inequality [149]

$$\frac{1}{t}\ln(1 + \frac{1}{xy}) \ge \frac{2}{\bar{t}} \ln\left(1 + \frac{1}{\bar{x}\bar{y}}\right) + \frac{2}{\bar{t}(\bar{x}\bar{y} + 1)}$$
$$- \frac{1}{\bar{t}\bar{x}(\bar{x}\bar{y} + 1)}x - \frac{1}{\bar{t}\bar{y}(\bar{x}\bar{y} + 1)}y - \frac{\ln(1 + 1/\bar{x}\bar{y})}{\bar{t}^2}t \tag{12.9}$$
$$\forall t > 0, \bar{t} > 0, x > 0, \bar{x} > 0, y > 0, \bar{y} > 0.$$

Algorithm 11 : Joint optimal harvesting time and power allocation problem (12.8)

1: **Initialisation:** Set feasible points $\theta^{(0)}$, $\mathbf{p}^{(0)}$, $\kappa = 0$ and $\phi^{(0)} = \sum_{n=1}^{N} \psi_n(\theta^{(0)}, \mathbf{p}^{(0)})/\vartheta(\theta^{(0)}, \mathbf{p}^{(0)})$. Set the tolerance $\varepsilon = 10^{-2}$.
2: **Repeat**
3:   Solve the (12.13) for the optimal solution $(\theta^{(\kappa+1)}, \mathbf{p}^{(\kappa+1)})$. Set $\phi^{(\kappa+1)} = \sum_{n=1}^{N} \psi_n(\theta^{(\kappa+1)}, \mathbf{p}^{(\kappa+1)})/\vartheta(\theta^{(\kappa+1)}, \mathbf{p}^{(\kappa+1)})$.
4: Set $\kappa := \kappa + 1$
5: **Stop** convergence of the objective in (12.13).

---

which follows from the convexity of function $\ln\left(1 + 1/xy\right)/t$.

For $x = p_n/h_{n,n}, y = \sum_{i \neq n}^{N} h_{i,n}/p_i + \sigma^2, t = \theta, \bar{x} = x^{(\kappa)} = p_n^{(\kappa)}/h_{n,n}$,
$\bar{y} = y^{(\kappa)} = \sum_{i \neq n}^{N} h_{i,n}/p_i^{(\kappa)} + \sigma^2, \bar{t} = t^{(\kappa)} = \theta^{(\kappa)}$, thus, the throughput can be approximated as

$$r_n(\theta, \mathbf{p}) \geq \psi_n^{(\kappa)}(\theta, \mathbf{p}) \tag{12.10}$$

where

$$\psi_n^{(\kappa)}(\theta, \mathbf{p}) = \frac{2}{t^{(\kappa)}} \ln\left(1 + \frac{1}{x^{(\kappa)}y^{(\kappa)}}\right) + \frac{2}{t^{(\kappa)}(x^{(\kappa)}y^{(\kappa)} + 1)} - \frac{1}{t^{(\kappa)}x^{(\kappa)}(x^{(\kappa)}y^{(\kappa)} + 1)}x -$$
$$\frac{1}{t^{(\kappa)}y^{(\kappa)}(x^{(\kappa)}y^{(\kappa)} + 1)}y - \frac{\ln(1 + 1/x^{(\kappa)}y^{(\kappa)})}{(t^{(\kappa)})^2}t. \tag{12.11}$$

With the feasible points $(\theta^{(k)}, \mathbf{p}^{(k)})$ of (12.8), one has

$$\phi^{(k)} = \sum_{n=1}^{N} \psi_n\left(\theta^{(k)}, \mathbf{p}^{(k)}\right) / \vartheta\left(\theta^{(k)}, \mathbf{p}^{(k)}\right). \tag{12.12}$$

At the kth iteration, the following convex program is solved to generate the next feasible point

$$\max_{\theta, \mathbf{p}} \sum_{n=1}^{N} \psi_n^{(\kappa)}(\theta, \mathbf{p}) - \phi^{(\kappa)} \vartheta^{(\kappa)}(\theta, \mathbf{p}) \quad \text{s.t. } (12.7), (12.8b), \tag{12.13a}$$

$$\psi_n^{(\kappa)}(\theta, \mathbf{p}) \geq \bar{r}, \; n \in \mathcal{N}. \tag{12.13b}$$

where $\vartheta^{(\kappa)}(\theta, \mathbf{p}) = \sum_{n=1}^{N} 1/(\theta p_n) + (1 - 2/\theta^{(\kappa)} + \theta/(\theta^{(\kappa)})^2)\eta P_0 + P_{\text{cir}}$.

We propose an algorithm to solve the EE maximisation problem (12.13). The initial point $(\theta^{(0)}, \mathbf{p}^{(0)})$ for (12.11) is easily located by random search such that it satisfies the constraints in the problem (12.8).

12.4 Near-optimal resource allocation algorithms for EE maximisation

In this section, two low-complexity procedures are presented as conventional methods to evaluate the effectiveness of joint harvesting time and power allocation (JHTPA) in EE performance and solving time.

12.4.1 Optimal power allocation

This algorithm addresses the power allocation for the EE maximisation problem (12.5) where the harvesting time value is fixed at $1 - \tau = 1/\theta_{\text{fix}}, \theta_{\text{fix}} > 1$. Thus, the problem (12.5) is equivalent to

$$\max_{\mathbf{p}} \phi = \frac{\sum_{n=1}^{N} r_n(\theta_{\text{fix}}, \mathbf{p})}{\vartheta(\theta_{\text{fix}}, \mathbf{p})} \tag{12.14a}$$

$$\text{s.t. } p_n \le (\theta_{\text{fix}} - 1)\eta P_0 g_n, \ n \in \mathcal{N} \tag{12.14b}$$

$$\ln\left(1 + \frac{p_n h_{n,n}}{\sum_{i\ne n}^N h_{n,i} p_i + \sigma^2}\right) \ge \theta_{\text{fix}}\bar{r}, \ n \in \mathcal{N}. \tag{12.14c}$$

where $\vartheta(\theta_{\text{fix}}, \mathbf{p}) = \sum_{n=1}^N p_n/\theta_{\text{fix}} + (1 - 1/\theta_{\text{fix}})\eta P_0 + P_{\text{cir}}$.

To solve the nonconvex problem (12.14), we apply the inequality (12.9) for $x = 1/p_n h_{n,n}$, $y = \sum_{i\ne n}^N h_{n,i} p_i + \sigma^2$, $t = 1$, and $\bar{x} = x^{(\kappa)} = 1/p_n^{(\kappa)} h_{n,n}$, $\bar{y} = y^{(\kappa)} = \sum_{i\ne n}^N h_{n,i} p_i^{(\kappa)} + \sigma^2$, $\bar{t} = t^{(\kappa)} = 1$.

Then, the numerator of the objective function in (12.14) can be approximated as

$$r_n(\theta_{\text{fix}}, \mathbf{p}) \ge \bar{\psi}_n^{(\kappa)}(\theta_{\text{fix}}, \mathbf{p}) \tag{12.15}$$

where $\bar{\psi}_n^{(\kappa)}(\theta_{\text{fix}}, \mathbf{p})$ is defined as (12.11).

At the κth iteration, the following convex program is solved to generate the next feasible point:

$$\max_{\mathbf{p}} \sum_{n=1}^N \bar{\psi}_n^{(\kappa)}(\theta_{\text{fix}}, \mathbf{p}) - \phi^{(\kappa)}\vartheta(\theta_{\text{fix}}, \mathbf{p}) \tag{12.16a}$$

$$\text{s.t. } (12.14b), (12.14c) \tag{12.16b}$$

where $\phi^{(\kappa)} = \sum_{n=1}^N \bar{\psi}_n(\theta_{\text{fix}}, \mathbf{p}^{(\kappa)})/\vartheta(\theta_{\text{fix}}, \mathbf{p}^{(\kappa)})$.

12.4.2   Optimal harvesting time

This algorithm solves the harvesting time optimisation problem in the slack variable of θ with the use of maximum harvested power in D2D communication as follows:

$$p_n = (\theta - 1)\eta P_0 g_n \tag{12.17}$$

Therefore, the max-min sum-rate problem with fixed harvested energy is given by

$$\max_{\theta} \min_{n\in\mathcal{N}} r_n(\theta) \quad \text{s.t. } (12.7) \tag{12.18}$$

where $r_n(\theta) = \frac{1}{\theta}\ln\left(1 + \frac{(\theta-1)h_{n,n}g_n}{(\theta-1)\sum_{i\ne n}^N h_{n,i}g_i + \sigma^2/\eta P_0}\right)$.

Next, the objective function in (12.18) can be approximated by using the inequality (12.9) for $x = 1/(\theta - 1)h_{n,n}g_n$, $y = (\theta - 1)\sum_{i\ne n}^N h_{i,n}g_i + \sigma^2/\eta P_0$, $t = \theta$, $\bar{x} = x^{(\kappa)} = 1/(\theta^{(k)} - 1)h_{n,n}g_n, \bar{y} = y^{(\kappa)} = (\theta^{(k)} - 1)\sum_{i\ne n}^N h_{i,n}g_i + \sigma^2/\eta P_0$, $\bar{t} = t^{(\kappa)} = \theta^{(\kappa)}$, and thus one has

$$r_n(\theta) \ge \hat{\psi}_n^{(\kappa)}(\theta) \tag{12.19}$$

where $\hat{\psi}_n^{(\kappa)}(\theta)$ is defined as (12.11).

At the κth iteration, the following max-min program is solved to generate the next feasible point

$$\max_{\theta} \min_{n\in\mathcal{N}} \hat{\psi}_n^{(\kappa)}(\theta) \quad \text{s.t. } (12.7). \tag{12.20}$$

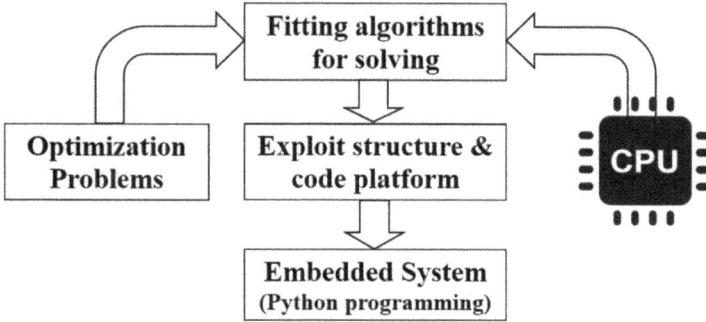

Figure 12.2   A structure of real-time embedded optimisation. ©IEEE 2018. Reprinted with permission from [259].

We assume that θ^* is the optimal solution of the problem (12.18). Then, the EE performance is defined as

$$\phi(\theta^*) = \frac{\sum_{n=1}^{N} r_n(\theta^*)}{\vartheta(\theta^*)} \tag{12.21}$$

where $\vartheta(\theta^*) = (1 - 1/\theta^*)\eta P_0(\sum_{n=1}^{N} g_n + 1) + P_{\text{cir}}$.

12.5   Implementation

In this section, we evaluate the performance of the UAV network by embedded optimisation module implemented in Python [223]. The results are obtained using CVXPY 0.4.11 package with ECOS solver. The computational platform is a laptop with an Intel® Core™ i7, CPU @2.80 GHz and 16 GB memory.

   Similar to [268], the channel power gain between D2D-Tx and D2D-Rx is modelled as

$$h_{n,n} = \beta_0 \rho_n^2 D^{-\alpha_h} \tag{12.22}$$

where β_0 is the channel power gain at the reference distance $d0$, ρ_n is an exponentially distributed random variable with unit mean, D is the distance between D2D-Tx and D2D-Rx and α_h represents the path loss exponent for D2D links.

   We exploit the air-to-ground (ATG) channel model for D2D UAV-assisted communication [270]. The channel power gain from the UAV to the nth D2D-Tx located at (x, y) under the LoS or NLoS links is given by

$$g_n(x, y) = Pr_{LoS} \times (\sqrt{x^2 + y^2 + H^2})^{-\alpha_g} + Pr_{NLoS} \times \gamma(\sqrt{x^2 + y^2 + H^2})^{-\alpha_g} \tag{12.23}$$

where $Pr_{LoS} = 1/(1 + a \times \exp(-b[\varphi - a]))$ is the LoS probability [271], where a and b are constant values depending on the environment. Then, one has $Pr_{NLoS} = 1 - Pr_{LoS}$,

Table 12.1   Simulation parameters

Parameter	Numerical value
Bandwidth	1 MHz
UAV transmission power	5 W
Path-loss exponent	$\alpha_h = 3, \alpha_g = 4$
Channel power gain at the reference	$\beta_0 = -30$ dB
Noise power density	-130 dBm/Hz
Energy harvesting efficiency	$\eta = 0.5$
UAV non-transmission power	4 W
ATG channel parameters	$a = 11.95, b = 0.136$
Excessive attenuation factor	$Y = 20$ dB

and α_g represents the path loss exponent from UAV to D2D-Tx. The elevation angle φ in terms of degree unit is given by $\varphi = \frac{180}{\pi} \times \sin^{-1}\left(\frac{H}{\sqrt{x^2+y^2+H^2}}\right)$.

The QoS constraint is set as

$$\bar{r} = \min\{r_n(\theta_{\text{fix}}), 0.2\}\text{bps/Hz} \tag{12.24}$$

where $r_n(\theta_{\text{fix}})$ is defined in (12.18).

Other simulation parameters in the considered D2D UAV network are provided in Table 12.1 [268].

Figure 12.3 plots the average running time for solving the algorithms of JHTPA, optimal power allocation (OPA) and optimal harvesting time (OHT). As can be observed from this figure, the solving time of all the algorithms is in milliseconds with 10 D2D pairs. For instance, with 5 D2D pairs, the running time is lower than 50 milliseconds for OPA and OHT and around 150 milliseconds for JHTPA.

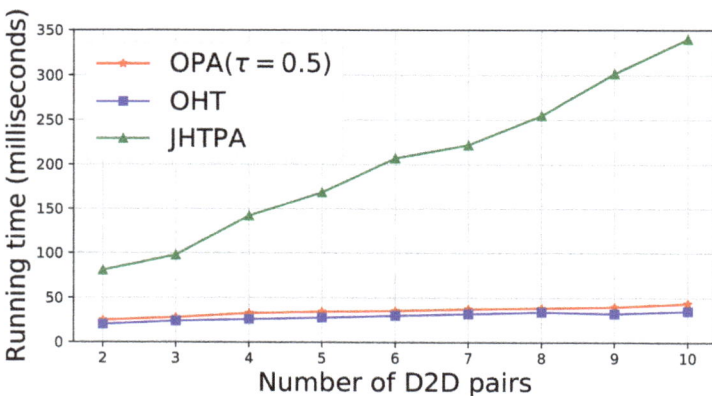

Figure 12.3   The average running times of OPA for $\tau = 0.5$, OHT and JHTPA algorithms versus the number of D2D pairs. ©IEEE 2018. Reprinted with permission from [263].

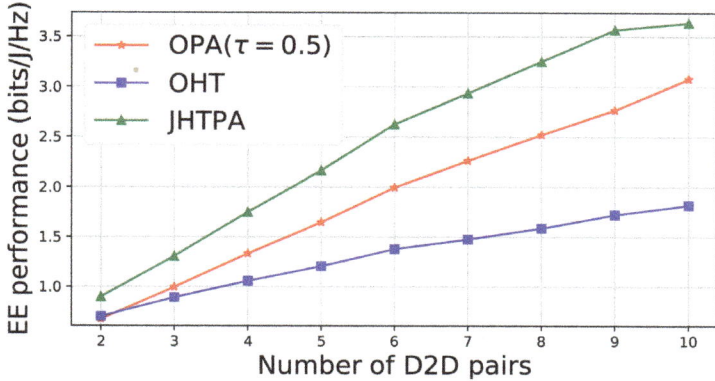

Figure 12.4   *The EE performance of OPA for τ = 0.5, OHT and JHTPA*
                  algorithms versus the number of D2D pairs. ©IEEE 2018.
                  Reprinted with permission from [263].

From Figures 12.3 and 12.4, we demonstrate the trade-off between the solving time and the EE performance, which should be carefully considered in real-time applications. Although the running time for JHTPA algorithm is higher than that for OPA and OHT, the EE performance in JHTPA significantly outperforms the two other algorithms. Interestingly, the EE performance for OPA algorithm is much higher than that for OHT algorithm, whereas the running time in these two algorithms is almost identical. As such, the OPA algorithm offers a better solution than the OHT algorithm.

12.6   Conclusion

In this chapter, we have proposed the real-time resource allocation for D2D communications assisted by a UAV. We have shown that our real-time optimisation is very suitable for UAV applications, where real-time control is crucial.

Chapter 13

Real-time deployment and resource allocation for distributed UAV systems in disaster relief

This chapter provides a robust and efficient resource allocation[a] for embedded UAV-enabled cellular networks in disaster communications. To recover network in disaster area, a fast user (UE) clustering based on K-means procedure and distributed control power coefficient will be proposed and can be embedded programming in the real system by using UAV-assisted relaying for real-time recovering and rescuing working network during and after disasters. The algorithms of low computational complexity with fast convergence are proposed for our expected solution. Numerical examples are provided to demonstrate the benefit of the proposed computational approaches.

13.1 Introduction

There is a lack of reliable support of existing technologies for data transmission in disaster-relief networks. In disaster communication, traditional cellular networks are often disrupted due to the lost connection between base stations (BSs) and UEs [273]. Under such extreme conditions, device-to-device networks can be applied (in combination with optimal route path algorithms) to maintain disaster communication network [274]. However, it is very challenging to select possible gateways or optimise routing in disaster areas. It is necessary to set up an emergency network when wireless access is restricted to UEs of a particular location in inhospitable disaster terrains. To this end, with flexible exploitation and mobility, unmanned aerial vehicles (UAVs)-based cellular networks have emerged as a promising solution to increase the reliability of wireless networks by serving as cellular-connected backbones [263]. By exploiting the line-of-sight (LoS) communications, UAVs can act as relay nodes to serve multiple UEs in disaster terrains who cannot be directly connected to the BS.

However, the deployment of UAV devices is limited by the stringent energy and deployment constraints. Moreover, the UAV access points need to reorganise and self-deploy to cope with the dynamically changing environment. Under disaster

---

[a] 1 This chapter has been as part of [272].

scenarios, self-organising the UAV network should be implemented in real-time [97, 263, 275]. Efficient UAV deployment has been recently considered to improve network performance [241, 259, 276, 277]. Nevertheless, there is a lack of practical UAV deployment and feasible resource management approaches considering real-time scenarios in critical-mission communication, i.e., natural disasters. In this chapter, we provide a framework of real-time optimisation for distributed relay-aided UAV systems in disaster scenarios based on the fast deployment of UAVs and efficient resource allocation with quality-of-service (QoS) requirement. In particular, we exploit a popular method of cluster selection model, called K-means [278], to form the cluster of UEs in the disaster area. Thus, a set of UE-specified pair-wise constraints on the connection between the UAVs and the UEs can be provided under constrained K-means clustering (CKC) procedure. The main contributions of this chapter are as follows:

- A central station deploys a number of UAVs as relay nodes to support a large number of UEs in a disaster area who cannot be served by the cellular BS due to poor channel conditions and high chance of lost connection. To this end, a fast and efficient UE clustering selection model based on K-means procedure is proposed for the embedded relay-aided UAV network.
- Under the challenges of limited spectrum, expensive resources and energy-efficient communication in UAV systems, a distributed real-time resource allocation will be proposed for maximising end-to-end (E2E) sum-rate and embedded programming in UAV devices for rapidly recovering the network to support a large number of UEs in disaster communications.

13.2   System model and problem formulation

13.2.1   System model

We consider the downlink transmission in a massive multiple-input multiple-output (MIMO) BS equipped with N antennas to serve some UEs in a safety area. Meanwhile, K UEs, who are randomly distributed in the disaster area, are grouped into M clusters. To support M clusters of UEs in the disaster coverage, we use M UAVs based on cellular connected to the BS, as shown in Figure 13.1. The UAVs and UEs are equipped with single antennas. Cluster m can serve a finite number of UEs with the set of $\mathcal{K}_m = \{1, ..., K_m\}$, $m = 1, .., M$. The UE (m, k) denotes the kth UE in cluster m.

We assume the 3D location of the BS, the UAVs and the UEs are (x_0, y_0, H_0) and $(x_{U,m}, y_{U,m}, H_{U,m})$, $m = 1, ..., M$ and (x_k, y_k), $k = 1, ..., K$, respectively. H_0 and $H_{U,m}$ are the antenna altitude of the BS and the UAV, respectively. We assume that the antenna altitude of a UAV is also its altitude. Then the distance between the BS and the mth UAV is expressed as

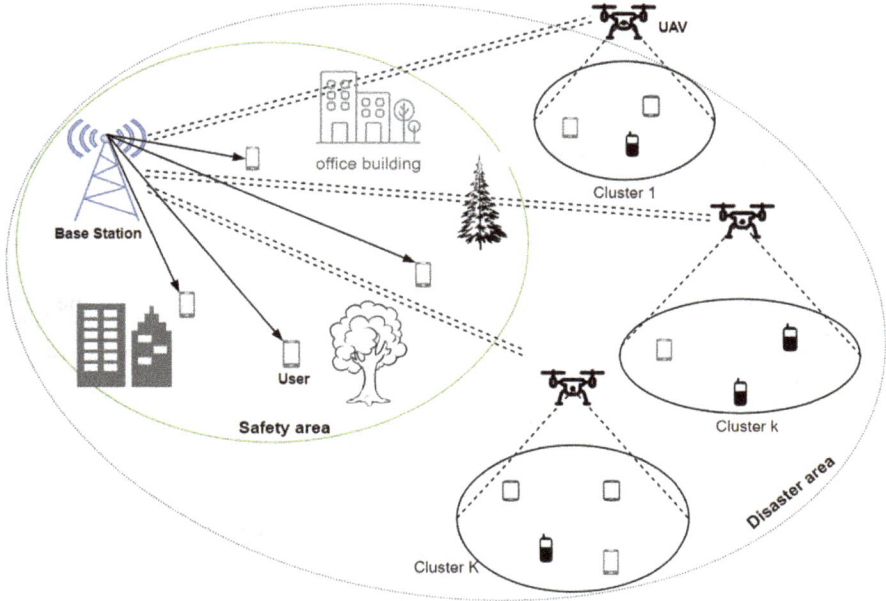

Figure 13.1   A typical model of cellular-connected UAV communication in disaster relief. ©IEEE 2018. Reprinted with permission from [272].

$$R_{0,m} = \sqrt{d_{0,m}^2 + (H_{U,m} - H_0)^2} \tag{13.1}$$

where $d_{0,m} = \sqrt{(x_{U,m} - x_0)^2 + (y_{U,m} - y_0)^2}$.

Similarly, the distance between UAV m and UE k in its cluster is given as

$$R_{m,k} = \sqrt{d_{m,k}^2 + H_{U,m}^2}, k \in \mathcal{K}_m \tag{13.2}$$

where $d_{m,k} = \sqrt{(x_{U,m} - x_k)^2 + (y_{U,m} - y_k)^2}$ is the Euclidean distance between UAV m and UE k.

Due to the LoS propagation and 3D location of the UAVs, the air-to-air (ATA) channel of the BS and the UAVs differs from the terrestrial channel. As such, the effect of the environment on LoS occurrence becomes critical. Hence, the path loss between the BS and UAV m is only based on free-space path loss model and given as

$$\beta_{0,m} = \beta_0 R_{0,m}^{-2} = \frac{\beta_0}{(x_{U,m} - x_0)^2 + (y_{U,m} - y_0)^2 + (H_{U,m} - H_0)^2} \tag{13.3}$$

where β_0 is the channel power gain at reference distance d_0.

The air-to-ground (ATG) channel between the UAVs and the UEs is more complex due to the effects of propagation blockage, e.g., the shadowing and blockage geometry. The path loss expression between UAV m and UE k is denoted as [241]

$$\beta_{m,k} = PL_{m,k} + \eta^{LoS} P_{m,k}^{LoS} + \eta^{NLoS} P_{m,k}^{NLoS} = 10\alpha \log(\sqrt{d_{m,k}^2 + H_{U,m}^2}) + AP_{m,k}^{LoS} + B$$

(13.4)

where η^{LoS} and η^{NLoS} are the average additional losses for LoS and NLoS, respectively, $A = \eta^{LoS} - \eta^{NLoS}$ and $B = 10\alpha \log(\frac{4\pi f_c R_{m,k}}{c}) + \eta^{NLoS}$. Therein, the distance path loss is given as

$$PL_{m,k} = 10 \log(\frac{4\pi f_c R_{m,k}}{c})^\alpha$$

(13.5)

where f_c is carrier frequency (Hz), c is the speed of light (m/s) and $\alpha \geq 2$ is the path loss exponent. The probability of LoS and NLoS is given by [242]

$$P_{m,k}^{LoS} = \frac{1}{1 + a \exp\left[-b\left(arc\tan\left(\frac{H_{U,m}}{d_{m,k}}\right) - a\right)\right]}$$

(13.6)

where a and b are constants depending on environment. Thus, one has $P_{m,k}^{NLoS} = 1 - P_{m,k}^{LoS}$. On the other hand, the small-scale fading of all channels $(\mathbf{h}_{0,m} \in \mathbb{C}^N, h_{m,k} \in \mathbb{C})$ is assumed as i.i.d. random variable with zero mean and unit variance.

We will exploit the structure of the ATA and ATG channels as $\mathbf{g}_{0,m} = \sqrt{\beta_{0,m}}\mathbf{h}_{0,m}$ and $g_{m,k} = \sqrt{\beta_{m,k}}h_{m,k}$, respectively.

We consider two phases of downlink transmission. In the first phase, the BS transmits signals to the UAVs. The signal received at UAV m is given by

$$y_{0,m} = \underbrace{\mathbf{g}_{0,m}^H \sqrt{P_0}\mathbf{f}_{0,m}x_{0,m}}_{\text{desired signal}} + \underbrace{\sum_{m' \in \mathcal{M}\backslash\{m\}} \mathbf{g}_{0,m'}^H \sqrt{P_0}\mathbf{f}_{0,m'}x_{0,m'}}_{\text{co-tier interference}} + n_m$$

(13.7)

where $\mathbf{g}_{0,m} \in \mathbb{C}^N$ is the channel between the BS and UAV m; $\mathbf{f}_{0,m}$ is the vector beamforming and $x_{0,m} \in \mathbb{C}$ is the information received at the mth UAV with $\|x_{0,m}\|^2 \leq 1$; P_0 is the transmit power at the BS; and $n_m \sim \mathcal{CN}(0, \sigma_m^2)$ is the additive white Gaussian noise (AWGN) at UAV m.

After the channel estimation process, the BS uses the estimated effective channels to design the precoding matrix. We employ the maximal ratio transmission (MRT), which is a simple and nearly optimal precoding design in massive MIMO networks [281]. The MRT downlink precoders at the BS and the UAV are given by

$$\mathbf{f}_{0,m} = \sqrt{p_{0,m}}\frac{\mathbf{g}_{0,m}^*}{\|\mathbf{g}_{0,m}\|}$$

(13.8)

where $p_{0,m}$ is power control coefficient.

In the second phase, the UAVs forward signals to the UEs in their cluster. The signal received by UE k in cluster m is written as

$$y_{m,k}\underbrace{g_{m,k}\sqrt{P_m}\sqrt{p_{m,k}}x_{m,k}}_{\text{desired signal}} + \underbrace{\sum_{l=1,l\neq m}^{M}\sum_{j\in\mathcal{K}_l} g_{l,km}\sqrt{P_l}\sqrt{p_{l,j}}x_{l,j}}_{\text{inter-cluster interference}} + n_{m,k}$$

(13.9)

where $g_{m,k} \in \mathbb{C}$ is the channel between UAV m and UE k, P_m is the transmit power at the mth UAV, $p_{m,k}$ and $x_{m,k}$ are the power control coefficient and information received at UE k by cluster m with $\|x_{m,k}\|^2 \le 1$ and $n_{m,k} \sim \mathcal{CN}(0, \sigma_{m,k}^2)$ is the AWGN. We assume that the intra-cluster interference in each cluster is cancelled by an inter-user interference suppression scheme [280].

For $\mathbf{p}_0 = [p_{0,m}]_{m=1}^{M}$ and $\mathbf{p}_m = [p_{m,k}]_{k=1}^{K_m}$, the network interference is characterised by the co-tier interference as

$$\mathcal{I}_m^{\text{cotier}}(\mathbf{p}_0) = \sum_{m' \in \mathcal{M} \backslash \{m\}} P_0 p_{0,m'} \|\mathbf{g}_{0,m}^H \mathbf{f}_{0,m'}\|^2, m \in \mathcal{M} \tag{13.10}$$

and the inter-cluster interference function[b] is written as

$$\mathcal{I}_{m,k}^{\text{inter}}(\mathbf{p}_m) = \sum_{l \in \mathcal{N}_m} \sum_{j \in \mathcal{K}_l} P_l p_{l,j} \beta_{l,k}, \ k \in \mathcal{K}_m; m \in \mathcal{M} \tag{13.11}$$

The information throughput in the first phase at the mth UAV (in nats) is

$$R_{0,m}(\mathbf{p}_0) = \frac{1}{2} B \ln \left(1 + \frac{P_0 p_{0,m} \|\mathbf{g}_{0,m}^H \mathbf{f}_{0,m}\|^2}{\mathcal{I}_m^{\text{cotier}}(\mathbf{p}_0) + \sigma_m^2} \right) \tag{13.12}$$

where B is the bandwidth of the system.

The information throughput in the second phase at the UE (m, k) (in nats) is

$$R_{m,k}(\mathbf{p}_m) = \frac{1}{2} B \ln \left(1 + \frac{P_m p_{m,k} \beta_{m,k} |h_{m,k}|^2}{\mathcal{I}_{m,k}^{\text{inter}}(\mathbf{p}_m) + \sigma_{m,k}^2} \right). \tag{13.13}$$

Then, the E2E information throughput at the UE (m, k) is given by

$$R_{m,k}^{E2E}(\mathbf{p}_0, \mathbf{p}_m) = \min\{R_{0,m}(\mathbf{p}_0), R_{m,k}(\mathbf{p}_m)\}. \tag{13.14}$$

Thus, the total E2E throughput of the mth cluster is written as

$$R_m(\mathbf{p}_0, \mathbf{p}_m) = \sum_{k \in \mathcal{K}_m} R_{m,k}^{E2E}(\mathbf{p}_0, \mathbf{p}_m). \tag{13.15}$$

13.2.2 Problem formulation

In this chapter, our target is to maximise the network throughput via UAVs clustering selection model and power allocation. To this end, the distributed maximin E2E rate problem is formulated as

$$\max_{\mathbf{p}_0, \mathbf{p}_m, m \in \mathcal{M}} \ \min_{k \in \mathcal{K}_m} \ R_{m,k}^{\text{E2E}}(\mathbf{p}_0, \mathbf{p}_m) \tag{13.16a}$$

---

[b]The intercell channel $h_{l,k}$ is difficult to estimate, and, thus, it must be defined as in (13.11).

$$\text{s.t.} \sum_{m \in \mathcal{M}} p_{0,m} \leq 1, \sum_{k \in \mathcal{K}_m} p_{m,k} \leq 1, m \in \mathcal{M} \tag{13.16b}$$

$$R_{m,k}^{E2E}(\mathbf{p}_0, \mathbf{p}_m) \geq \bar{r}_{m,k}, \ m \in \mathcal{M} \tag{13.16c}$$

$$(m, k) \in \mathcal{K}_m, m \in \mathcal{M} \tag{13.16d}$$

where (13.16b) represents the power constraints at the BS and the UAVs, the constraint (13.16c) sets the QoS data rate requirement at the first phase and the second phase and (13.16d) is the cluster selection model. Obviously, the problem (13.16) is nonconvex with the nonconvex functions (13.16c) and (13.16d).

To efficiently solve the above nonconvex problem, we separate the problem (13.16) into two subproblems. First, the UE clustering method with constrained QoS will be proposed for responding to constraints (13.16c) and (13.16d) by the CKC clustering procedure. Then, the distributed power allocation (Dist_OPA) is implemented to maximise the minimum E2E rate subproblems with constrained power budget.

13.3   Constrained K-means clustering method

13.3.1   *Preliminaries of K-means method*

In this chapter, the requirements of K-means clustering are based on large-scale path loss via devices' locations from global positioning system data such as the input data points $\mathbf{q}_k = (x_k, y_k), k = 1, ..., K$ (UE's location) and $\mathbf{q}_m = (x_{U,m}, y_{U,m}, H_{U,m})$, $m = 1, ..., M$ (UAV's location). Then, based on the cluster selection, the 3D locations of the UAVs will be chosen by the centroids $\mathbf{the}_m = (x_m, y_m, z_m), m = 1, ..., M$.

However, the conventional K-means clustering approach may not be suitable for UAVs with a limitation of QoS requirement. In this section, we propose a CKC method corresponding to the QoS constraints.

13.3.2   *Clustering model with QoS constraints*

Constrained UE clustering is a useful way to express a priori knowledge as to which UEs should or should not be grouped together. Thus, we provide two types of pairwise constraints as

- Must-link constraints $(m, k) \in \mathcal{C}_{\text{must}}$ indicate that the kth UE has to be located in cluster m with QoS constraints satisfied.
- Cannot-link constraints $(m, k') \in \mathcal{C}_{\text{not}}$ imply that the kth UE should not be placed in cluster m.

Let γ_{QoS} be the path loss threshold corresponding to the QoS requirement [281]. Then, a set of must-link pairs represents the satisfied QoS constraints such that the UE k is served by the UAV m and $\beta_{m,k} \leq \gamma_{QoS}$, i.e.,

$$d_{m,k}^2 + H_{U,m}^2 \leq 10^{\frac{\gamma_{QoS} - (AP_{m,k}^{LoS} + B)}{10\alpha}}. \tag{13.17}$$

In contrast, a set of cannot-link pairs represents the violation of QoS constraints. In Algorithm 12, we propose a CKC algorithm.

13.3.3 Selecting the number of clusters

In this chapter, the number of clusters is estimated by Algorithm 12, which corresponds to the number of UAVs. However, the number of UAVs, which can be provided by the system, has to be managed and deployed to recover the disaster network under QoS requirement. For energy efficiency, the selected number of clusters (UAVs) should be as small as possible and must be smaller than the maximum number of UAVs provided by the system. Since the K-means algorithms are unsupervised clustering, finding the suitable number of clusters plays a role in clustering analysis.

Algorithm 12 Constrained clustering based on K-means

1: **Initialisation**: The UAVs' locations $(\mathbf{q}_m)$ are randomly initialised as the centroid $\{\boldsymbol{\theta}_m^{(0)}\}, m = 1, ..., M$. The maximum number of iterations is set at N_{max}.

2: **Repeat**

3:    **Update index set of users:**

4:    Compute the distance $\text{dist}(\mathbf{q}_k, \boldsymbol{\theta}_m), m = 1, .., M$. Then, assign appropriate users into their cluster with the smallest distance.

5:    Repeat step (4) until all users have been labelled.

6:    **Update UAV's altitude:**

7:    Compute path loss between the UAV and their users.

8:    If QoS constraints are not violated. Go to next step.

9:    If QoS constraints are violated. Update the altitude of the UAV to satisfy constraints (13.17) for the worst user. Else return with the best of service.

10:    **Update centroids:**

11:    Update the centroid location for each cluster as

12:    $\boldsymbol{\theta}_m = \frac{1}{K_m} \sum_{k \in \mathcal{K}_m} \mathbf{q}_k$

13: **Until** The cluster members do not change or the procedure reaches to N_{max}.

13.4 Maximising end-to-end throughput via distributed power allocation

In this section, we provide a distributed resource allocation algorithm for solving the worst E2E rate maximisation problem (13.16). This approach is efficiently implemented by applying Block Coordinate Descent (BCD) procedure.

For improving the effective network system, the BS and the UAVs are cooperating for exchanging the UEs' information and power allocation. First, the transmit power of the BS and the UAVs is randomly initialised as $\{\mathbf{p}_0^{(0)}, \mathbf{p}_m^{(0)}\}, m \in \mathcal{M}$. For cluster m, a power allocation scheme is proposed in order to optimise the variables of power coefficients in $\mathbf{p}_m$. Meanwhile, the control power coefficients of other clusters are fixed. To this end, the two vectors of power control coefficients $\mathbf{p}_{0,[m:]} = [\bar{p}_{0,1}, ..., \bar{p}_{0,m-1}, p_{0,m}, \bar{p}_{0,m+1}, ..., \bar{p}_{0,M}]^T$ and $\mathbf{p}_{[m:]} = [\bar{\mathbf{p}}_1, ..., \bar{\mathbf{p}}_{m-1}, \mathbf{p}_m, \bar{\mathbf{p}}_{m+1}, ..., \bar{\mathbf{p}}_M]^T$ are received at the UAV m via the cooperative network, where $\bar{p}_{0,i}$ and $\bar{\mathbf{p}}_i, i \neq m$ are fixed. For each cluster, the maximin E2E rate subproblem under power budgets is shown as

$$\max_{\mathbf{p}_0, \mathbf{p}_m, m \in \mathcal{M}} \min_{k \in \mathcal{K}_m} R_{m,k}^{E2E}(\mathbf{p}_0, \mathbf{p}_m) \tag{13.18a}$$

$$s.t. \ \|\mathbf{p}_{0,[m:]}\|_1 \leq 1, \ \|\mathbf{p}_m\|_1 \leq 1 \tag{13.18b}$$

This step is implemented for all the clusters. Then, the BCD procedure terminates with the convergence of the power allocation of all the clusters. Note that the subproblems (13.18) are convex with a concave objective function and convex constraints. Thus, these problems can be solved in a decentralised manner by using optimisation tools, i.e., CVX [221].

Consequently, Algorithm 13 summarises the proposed distributed power allocation (Dist_OPA) approach via the BCD procedure for the worst E2E rate maximisation problem.

Algorithm 13 Algorithm for distributed power allocation (13.18)

1: **Input:** Group of the clusters $\mathcal{M}$ by Algorithm 12. Set a feasible initial power $(\mathbf{p}_0^{(0)}, \mathbf{p}_m^{(0)})$. Set $\kappa = 0$ and the maximum number of iterations N_{max}.

2: **Output:** Optimal power allocation $\{\mathbf{p}_0^*, \mathbf{p}_m^*\}$.

3: **Repeat:**

4: **for** $m = 1$ to M **do**

5:     Solve subproblem (13.18) for cluster m (in parallel).

6:     **return** $\{\mathbf{p}_0^*, \mathbf{p}_m^*\}$ as the next entry of $\mathbf{p}_{0,[m:]}$ and $\mathbf{p}_{[m:]}$ for the next iteration.

7: **end for**

8: Update the vector power allocation for all clusters.

9: Set $\kappa := \kappa + 1$.

10: $(\mathbf{p}_0^*, \mathbf{p}_m^*)$ when the procedure converges or it reaches the maximum number of iterations.

13.5 Simulation results

In this section, the performance of the considered network is evaluated with the use of embedded optimisation programming, i.e., the CVX version 2.1 in MATLAB®

[221]. The computational platform is used for performing in a PC with AMD Ryzen 7 2700X, CPU @3.7 GHz and 32 GB memory.

For simulation, we set the system parameters as follows:

- The safety area is a circle coverage with a radius of 500 m.
- The disaster area is extended from the safety area with a radius up to 2 000 m.
- The location of the BS is (0, 0, 30) whereas the locations of UEs and UAVs are randomly distributed in the disaster area.
- The number of UEs is set as $K = \{50, 100, 200\}$ and the maximum number of UAVs is $M = \{10, 30, 60\}$.
- The limited altitude of the UAVs $(H_{U,min}, H_{U,max})$ is $(50, 200)$ m.
- The path loss threshold is set as $\gamma_{QoS} = 110$ dB.
- The tolerance and the maximum number of iterations for convergence of algorithms are $\varepsilon = 10^{-3}$ and $N_{max} = 20$.
- The bandwidth and carrier frequency are $B = 1$ MHz/ $f_c = 2$ GHz.
- The BS transmit power is set as 40 W.

Other parameters of the channel model are set as in [148, 281].

A total convergence of the proposed Algorithm 12 (CKC) for UE clustering and Algorithm 13 optimal power allocation (OPA) for power allocation is estimated versus several scenarios, where all of them are seen to converge rapidly within a few iterations. For instance, the number of iterations is $6, 7, 8$ for the number of UEs–UAVs $K = 50, M = 10$, $K = 100, M = 25$ and $K = 200, M = 48$, respectively.

Next, we show the running time of Algorithm 12 for CKC method. The figure demonstrates the fast deployment of the UAVs with K-means clustering. For example, the system requires 23 ms, 33 ms and 45 ms for solving clustering problem by Algorithm 12 with $(K = 50, M = 10)$, $(K = 100, M = 25)$ and $(K = 200, M = 48)$, respectively.

In our simulations, OPA refers to the optimal power coefficient allocation by the proposed Algorithms 12–13, while EPA refers to the equal-power allocation by adjusting Algorithm 12–13 for $[p_{0,m}] = p_0$ and $[p_{m,k}] = p_m$, $\forall k, m$. Also, fixed_EPA refers to the EPA approach under fixed power coefficient values such as $p_{0,m} = 1/M$ and $[p_{m,k}]_{k \in \mathcal{K}_m} = 1/K_m$, $\forall m$.

Figures 13.2 and 13.3 plot the sum worst E2E throughput performance of our proposed approaches versus the power-to-noise ratio at the UAVs. In these results, the number of UAVs used is increased for supporting an increase in the number of UEs. This set-up is to guarantee the ability of the system to provide coverage network. Obviously, the worst E2E rate of OPA significantly outperforms EPA and fixed_EPA. Interestingly, fixed_EPA provides a better worst E2E rate performance than EPA. This is because the EPA only optimises the transmit power of the UAVs without the control of power coefficients while they are controlled in fixed_EPA with their fixed values.

Next, the computation time for OPA, EPA and fixed EPA under centralised and distributed approaches is shown in Table 13.1. By using distributed computing, the relaying UAV-assisted system can save a lot of execution time compared

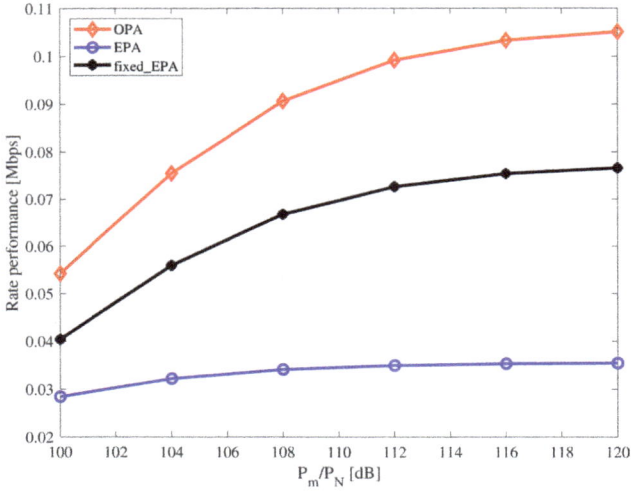

*Figure 13.2   The sum worst E2E throughput for OPA, EPA and fixed_EPA
schemes versus the power-to-noise ratio (P_m/P_N) of the UAVs under
$K = 50$ and $M = 10$. ©IEEE 2018. Reprinted with permission from
[272].*

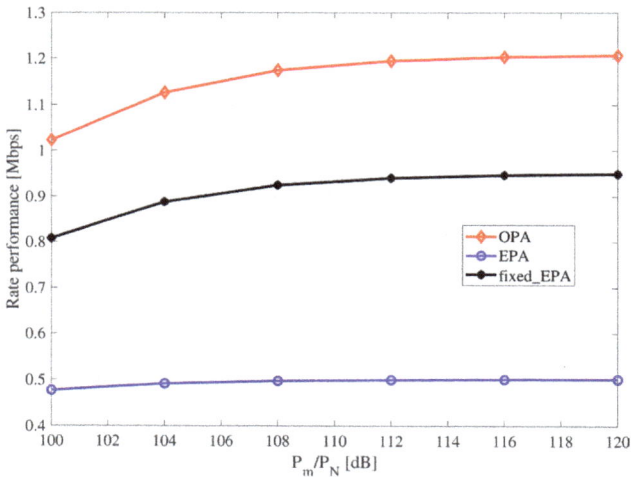

*Figure 13.3   The sum worst E2E throughput for OPA, EPA and fixed_EPA
schemes versus the power-to-noise ratio (45) of the UAVs under
$K = 200$ and $M = 48$. ©IEEE 2018. Reprinted with permission from
[272].*

Table 13.1 *Comparison of execution time for centralised/distributed OPA and*
EPA, and fixed EPA methods

Methods	# of UEs and UAVs		
	$K=50, M=10$	$K=100, M=25$	$K=200, M=48$
Cent_OPA (s)	15.89	37.02	71.45
Dist_OPA (s)	1.61	1.27	1.23
Cent_EPA (s)	23.10	54.16	104
Dist_EPA (s)	2.33	1.84	1.78
Fixed_EPA (ms)	26	42	71

to centralised computing. For $K = 100$ and $M = 25$, the computing time in distributed manner speeds up more than 27 times as in centralised manner. Although the fixed_EPA requires a small amount of computation time, the worst E2E rate of the fixed_EPA has significantly shorter computation time than OPA. As expected, the distributed OPA can satisfy both fast implementation and high-QoS for deploying the UAV relay networks in disaster relief.

On the other hand, we also provide the best service of the system by an unlimited number of UAVs. The best service is defined by satisfying the QoS requirement via path loss threshold (γ_{QoS}). As shown in Table 13.2, the number of 80 and 160 UEs will be guaranteed the QoS service with the average number of UAVs used at 37 and 47, respectively.

13.6 Conclusions

We have proposed the clustering selection and resource allocation algorithms of low-computational complexity with fast convergence for maximising the E2E sum-rate in the downlink transmission of a massive MIMO BS and UAV-cellular relaying networks while meeting real-time service and QoS constraints. The numerical results have demonstrated the advantage of the proposed approaches over the conventional schemes. This chapter has considered a novel trend of real-time optimisation for wireless communication systems under natural disasters. This approach will continue to see its adoption in the future advanced telecommunications networks in terms of coping with major issues of global challenges, especially with the global warming effect.

Table 13.2 *The best service for UAV-assisted relaying*

Number of UEs (K)	40	80	120	160	200
Average number of UAVs (M)	27	37	43	47	50

Chapter 14

Practical optimisation of path planning and completion time of data collection for UAV-enabled disaster communications

This chapter proposes efficient optimisation methods[a] for embedded relay-assisted unmanned ariel vehicles (UAVs) in wireless sensor networks (WSNs) to cope with the hazardous effect of natural disaster. Particularly, by using advanced optimisation techniques, proposed low-complexity procedures are suitably applied to internet-of-things (IoT) applications when the execution time is strictly governed in disaster scenarios. Our model considers real-time optimisation in embedded UAV-WSN communication for tracking and gathering sensor data. The algorithms have low computational complexity with fast deployment and low execution time for solving our problem in milliseconds. Numerical results are shown to demonstrate the benefit of our proposed approaches for UAV-WSN.

14.1 Introduction

Natural disasters often cause severe disruption to communication networks and power supply systems, which makes the prompt restoration and sustainability of communication networks a formidable task. Recently, UAVs have been considered as a promising approach in providing seamless connectivity by serving as relays (i.e., flying base stations) to provide the connection between the disaster area and control stations.

Utilising the dominant presence of line-of-sight (LoS) connections [228], UAV communication networks can be more efficient and inexpensive for the deployment of the IoT wireless network applications [263]. It is because UAVs can provide a better service for more wireless devices by enhancing network coverage. Hence, there are many types of applications in which UAVs can be exploited such as those in environmental remediation, navigation to gather data, or performing dangerous tasks in disaster relief [97, 232, 259]. For instance, UAVs can be deployed as relay nodes for tracking devices. They will need to collaborate with the central system

---

[a]This chapter has been as part of [277].

(ground station) in path planning and resource allocation. For real-time applications, the tracking results will be collected at the UAVs and reported to the central system within given strict time deadlines [259, 282].

Nevertheless, there are many issues of the UAV implementation that require a careful consideration, such as the efficient deployment and operation in UAV communications to provide high capability and capacity. In this chapter, we look into UAVs' trajectory planning and data gathering in a WSN during disaster recovery.

The number of UAVs depends heavily on the WSN scenarios, communication protocols, onboard processing power, and path planning models. A UAV may be present and implement its own modelling and operating system in complex procedures. The modelling of UAVs has opened many challenges and is an interesting research direction in many areas such as nonlinear system, nonconvex optimisation problem and constrained clustering model in UAV networks [232, 283]. K-means clustering is a popular method of cluster selection model with simple and fast implementation; however, this method may not be suitable for UAVs with many constraints and limitations. In this chapter, we propose a constrained K-means clustering method corresponding to real UAV constraints.

On the other hand, UAV devices have limited energy storage, e.g., current UAVs may only have enough power for a flight of minutes or maximum an hour in one charge. More importantly, in fast time-varying environments, the UAV nodes would dynamically work by frequently reorganising the network. This means their path planning needs to change over time and uses more energy to prolong the stability of the network. Therefore, the deployment and mission completion time issues should be considered in UAV-based applications.

14.2   UAV-WSN system model

14.2.1   System model

Multiple UAVs are used for tracking targets and collecting data from wireless sensors. In this chapter, the trajectory design of multiple UAVs with the set of $\mathcal{M} = \{1, ..., M\}$ is optimised and controlled by a ground station for tracking and gathering data from the sensor nodes, as provided in Figure 14.1. The target area $\mathcal{A}(\ell, d)$ is a rectangle with length ℓ and width d. There are $K \gg 1$ static ground sensors (GSs) spread and distributed between an initial point and final point over the considered network area, with K being denoted by the set $\mathcal{K} = \{1, ..., K\}$. The location of a GS $\mathbf{q}_k = [x_k, y_k, 0]^T, k \in \mathcal{K}$ is identified at the ground station by global positioning system (GPS). Each UAV is equipped with a single antenna.

We consider that all the UAVs operate over a duration $T > 0$ in seconds (s). The time-varying 3D location of the UAV at a time instant t, $0 \le t \le T$ is denoted as $\mathbf{q}_m(t) = [x_m(t), y_m(t), h_m(t)]^T \in \mathbb{R}^3, m \in \mathcal{M}$. Without loss of generality, the time length T is discretised into N equal time slots denoted by the set $\mathcal{N} = \{1, ..., N\}$. Therefore, the choice of the elemental time length for a typical slot is assumed as $\eta = T/N$, which is sufficiently small such that the location of the UAVs is stable

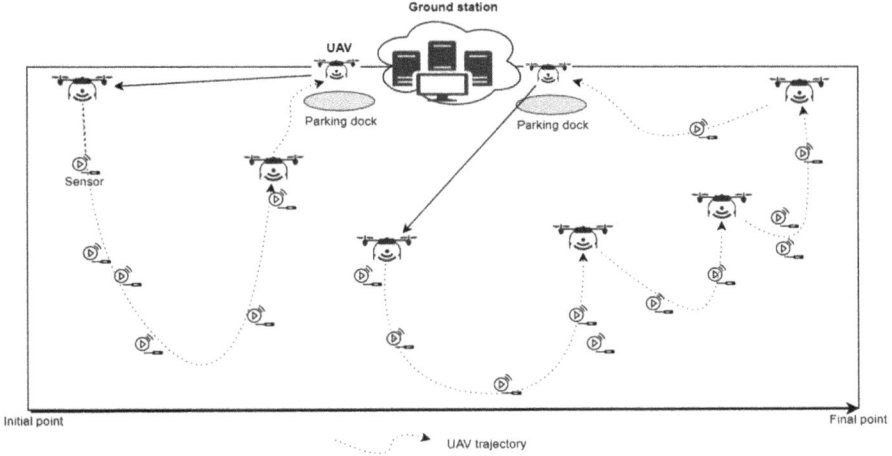

*Figure 14.1   A path planning UAV model for data collection in a WSN. ©IEEE
2018. Reprinted with permission from [277].*

within a time slot [284]. Thus, the trajectory of the UAV can be approximated as$q_m[n] = [x_m[n], y_m[n], h_m[n]]^T, n \in \mathcal{N}.$

To implement the mission, the UAVs fly from their parking dock $\mathbf{q}_{G,m} = [x_{G,m}, y_{G,m}, h_{G,m}]^T$, i.e., $\mathbf{q}_m[1] = \mathbf{q}_{G,m}$, follow their trajectory to collect sensed data from GSs and finish at their parking dock $\mathbf{q}_m[N] = \mathbf{q}_{G,m}$. In path planning, the trajectory of the UAVs is assumed to be in the considered area $\mathcal{A}$ while they are tracking and gathering sensor data. The gathered data packets of GSs will be transferred to the ground station at the end of the UAVs' trajectory.

We consider the air-to-ground channel model with LoS-dominated links for the wireless communication channel between the UAVs and GSs [285]. At the nth time slot of the mth UAV, the distance from the mth UAV and kth GS is shown as

$$R_{mk}[n] = \sqrt{D_{mk}^2[n] + h_m^2[n]} \tag{14.1}$$

where $D_{mk}[n] = \sqrt{(x_m[n] - x_k)^2 + (y_m[n] - y_k)^2}$ is the Euclidean distance between the UAV-GS (m, k).

Hence, the channel power gain of the UAV-GS (m, k) is denoted as

$$g_{mk}[n] = \beta_0 R_{mk}^{-2}[n] = \frac{\beta_0}{D_{mk}^2[n] + h_m^2[n]} \tag{14.2}$$

where β_0 is the channel power gain at the reference distance. On the other hand, the small-scale fading channels ($h_{mk} \in \mathcal{C}$) are neglected since they have little impact on LoS-dominated communication.

The flying time of the mth UAV between two adjacent time slots is as follows:

$$\delta_{m,n} = \frac{\|\mathbf{q}_m[n+1] - \mathbf{q}_m[n]\|}{V} \, , \, n = 1, ..., N \tag{14.3}$$

where V is the average flight speed of the UAVs.

To guarantee the domination of LoS links, we consider the coverage region corresponding to the quality-of-service (QoS) requirement by defining a circular disc with radius D_{cov}. The relation between the radius $D_{m,cov}$ and the altitude of UAV m is as follows:

$$h_m[n] = D_{m,cov} \tan(\theta), \, \forall n, m \tag{14.4}$$

where θ is set at 20.34° for suburban environment [281]. Therefore, one UAV only communicates to and gathers from the GSs in its coverage with the Euclidean distance of UAV-GS of less than the coverage region $D_{m,cov}$. The considered inequality is given as

$$D_{mk}[n] \leq D_{m,cov} \, , \, \forall n. \tag{14.5}$$

For a constant transmit power (P_k) at the kth GS, the uplink data rate at mth UAV is given as

$$C_{mk}[n] = B\log_2(1 + g_{mk}[n]P_k/\sigma_m^2) \tag{14.6}$$

where B is the bandwidth and σ_m^2 is the Gaussian noise power at UAV m.

14.2.2 Sensing data

We assume that the GSs forward their data packet, with the size of $\bar{D}_k$ using a standard data rate $\bar{C}_k$ to the UAVs. The standard transmission rate $\bar{C}_k$ follows a type of wireless IoT connectivity, i.e., low-power Bluetooth or LoRaWan for sensor networks [286]. Based on (15.20) and (14.6), the QoS requirement is written as

$$C_{mk}[n] \geq \bar{C}_k, \forall n, k, m. \tag{14.7}$$

or

$$h_m[n] \leq \sqrt{\frac{\beta_0 P_k}{\sigma_m^2(2^{\bar{C}_k/B} - 1)} \frac{1}{\tan^{-2}(\theta) + 1}} \, , \, \forall n, k, m. \tag{14.8}$$

The flying time of the UAVs over K GSs for the UAV-GS exchange duration is calculated as

$$L_{tot} = L_{con} + L_{data} + L_{tran} = \sum_{k \in \mathcal{K}} L_{k,con} + \sum_{k \in \mathcal{K}} \frac{\bar{D}_k}{\bar{C}_k} + \sum_{m \in \mathcal{M}} \sum_{n=1}^{N} \delta_{m,n} \tag{14.9}$$

where L_{con} and L_{data} are the connecting time and the data transferring time for the UAV-GS exchange duration. L_{tran} is the flying time of trajectory planning of the UAVs.

Thus, the minimum number of time slots to collect the data sensed at the K GSs at the transmission rate $\bar{C}_k$ is calculated as

$$\bar{N}_{\min} = \left\lceil \frac{L_{con} + L_{data} + L_{tran}}{\eta} \right\rceil. \tag{14.10}$$

On the other hand, the minimum number of UAVs necessary for the considered WSN is estimated as

$$M \geq \bar{M}_{min} = \left\lceil \frac{\bar{N}_{min}}{N} \right\rceil. \tag{14.11}$$

To guarantee the operating time of the UAVs in practice, a constraint of allowable flight time is written by

$$L_{tot} \leq MT_{max} \tag{14.12}$$

where T_{max} is a UAV's maximum operating time. For large-scale scenarios, the value of L_{tot} can be significantly higher than the period time T_{max}.

14.2.3 Problem formulation

Our target is to minimise the total flying time that the UAVs need to spend for collecting the data from K GSs. To this end, the minimisation of completion time under the minimum number of UAVs and their trajectory constraints is expressed as

$$\min_{M, \mathbf{q}_m} L_{tot} \tag{14.13a}$$
$$\text{s.t.} (15.20), (14.8), (14.11), (14.12) \tag{14.13b}$$
$$\mathbf{q}_{min} \leq \mathbf{q}_m \leq \mathbf{q}_{max}, \ m = 1, ..., M. \tag{14.13c}$$

where (14.13c) represents the trajectory constraints of the UAVs. Obviously, problem (14.13) is a nonconvex problem with nonconvex functions. Moreover, for large-scale scenarios, problem (14.13) is very complex when the number of time slots (N), the number of UAVs (M) and the number of GSs (K) are investigated as large numbers.

We propose two approaches for solving the above nonconvex problem (14.13) in real-time scenarios.

14.3 Optimal completion time by peer-to-peer UAV-GS networks

In this approach, we assume that the UAVs operate at the constant altitude $h_m = \bar{H}$, $\forall m$. The total time slot is set as $K + 2$ corresponding to K GSs together with an initial time slot and a final time slot for each UAV. We also assume that the trajectory of the UAVs at the initial and final points are $\mathbf{q}_m[0] = \mathbf{q}_m[K_m + 1] = \mathbf{q}_{G,m}$.

14.3.1 Estimating the number of UAVs

Instead of (14.3), the estimation of UAV's flying time in the path planning for K time slots is given as

$$\delta_{p2p} = \sum_{k=1}^{K-1} \delta_{m,k}^{p2p} = \sum_{k=1}^{K-1} \frac{\|\mathbf{q}_m[k+1] - \mathbf{q}_m[k]\|}{V} \tag{14.14}$$

where $\mathbf{q}_m[k] = \mathbf{q}_k$.

In Algorithm 14, we propose an efficient procedure for estimating the number of UAVs for the peer-to-peer (P2P) UAV-WSN.

Algorithm 14 Estimating the number of UAVs ($\bar{M}_{p2p}$)

1: **Input:** The initial UAVs's locations ($\mathbf{q}_m[0]$) and the GSs' location ($\mathbf{q}_k$).
2: Set $\mathcal{M}_{p2p} = \varnothing$. Set $m = 1$, $\mathcal{K}_m = \varnothing$ and $L_m^{p2p} = 0$.
3: **for** $k = 1$ to K **do**
4:    **if** $L_m^{p2p} + \delta_m^{st} + \delta_m^{ft} \leq T_{\max}$
5:        Calculate $L_m^{p2p} = L_m^{p2p} + L_{k,con}^{max} + \bar{D}_k/\bar{C}_k + \delta_{k,k}^{p2p}$.
6:        Add $\mathcal{K}_m = \{k\}$.
7:    **end if**
8:    $\mathcal{K}_m$ and $\mathcal{M}_{p2p} = \{m\}$.
9:        Set $m = m + 1$, $\mathcal{K}_m = \varnothing$ and $L_m^{p2p} = 0$.
10: **end for**
11: **output:** $\{\mathcal{M}_{p2p}\}$ and $\{\mathcal{K}_m = \{1, ..., K_m\}\}$, $\forall m$.

In Algorithm 14, the initial and final flying time of the mth UAV are expressed as

$$\delta_m^{st} = \frac{\|\mathbf{q}_m[0] - \mathbf{q}_k|_{k=1}\|}{V} \tag{14.15}$$

$$\delta_m^{fi} = \frac{\|\mathbf{q}_m[K_m + 1] - \mathbf{q}_k|_{k=K_m}\|}{V} \tag{14.16}$$

Thus, the total completion time in the P2P scenario is given as

$$L_{tot}^{p2p} = \sum_{m \in \mathcal{M}_{p2p}} L_m^{p2p} \tag{14.17}$$

where

$$L_m^{p2p} = \delta_m^{st} + \delta_m^{fi} + \sum_{k \in \mathcal{K}_m} \left(L_{k,con}^{max} + \frac{\bar{D}_k}{\bar{C}_k} + \delta_{m,k}^{p2p} \right) \tag{14.18}$$

such that

$$L_m^{p2p} \leq T_{\max}. \tag{14.19}$$

14.3.2 Proposed optimisation problem and solving approach

The subproblem of the completion time minimisation problem (14.13) without constraints of UAV number is expressed as

$$\min_{\mathbf{q}_m} \sum_{m \in \mathcal{M}_{p2p}} L_m^{p2p}(\mathbf{q}_m) \tag{14.20a}$$

$$\text{s.t.} (15.20), (14.8), (14.13c) \tag{14.20b}$$

$$L_{tot}^{p2p} \leq \bar{M}_{p2p} T_{\max}. \tag{14.20c}$$

Note that subproblem (14.20) is a low-complexity convex problem with the convex objective function and convex constraints compared to nonconvex problem (14.13). Thus, the above problem can be solved by using optimisation tools, i.e., CVX [221].

14.4 Optimal completion time by clustering UAV-GS networks (CUN)

In this section, we separate (14.13) into two subproblems. First, the GS clustering method with constrained QoS (14.7) will be proposed for grouping GSs and estimating the number of UAVs used by the constrained K-means clustering procedure. Then, a fast trajectory design is implemented to minimise the completion time of the UAV-WSN.

14.4.1 Constrained K-means clustering model

By following the coverage communication between $\bar{M}_{clus}$ UAVs and K GSs, all GSs are divided into multiple groups $\mathcal{K}_j = \{1, ..., K_j\}, K_j \geq 1, j = 1, ..., J, \bar{M}_{clus} \leq J$ such that $\sum_{j=1}^{J} K_j = K$ and $\mathcal{K} = \{\mathcal{K}_1, ..., \mathcal{K}_J\}$. Thus, one or more groups will transfer their data to the UAV under the constraints (14.7) and (14.12).

In Algorithm 15, we propose a constrained K-means clustering algorithm.

Algorithm 15 Constrained GS clustering based on K-means

1: **Input:** The GSs' location (q_k). The locations of the centroids of the clusters are initialised as $\{\theta_j = q_k\}, J = K$. The maximum number of iterations is set at N_{max}.
2: Set $J^{real} = \varnothing$ and $\mathcal{K} = \{1, ..., K\}$.
3: **Repeat**
4:   **Update index of clusters:**
5:     Compute the distance $\text{dist}(q_k, \theta_j), \forall j, k$.
6:     Set $j = 0$.
7:     **while** $j \leq J$ **do**
8:       Set $j = j + 1, \hat{k} = 1$.
9:       **for** $k = \hat{k}$ to K **do**
10:        **if** $\text{dist}(q_k, \theta_j) \leq D_{j,cov}$ are satisfied, thus, assigning the GS $\hat{k}$ into cluster j. Set $\mathcal{K}_j = \{k\}$.
11:        **else** break;
12:        **end if**
13:      **end for**
14:      Set $J^{real} = \{j\}$
15:    **end while**
16:   **Update the number of GSs in clusters:**
17:     Calculate the completing time (L_j^{clus}) in each group.
18:     If (14.24) is satisfied. Go to next step.
19:     If (14.24) is violated. Return with the re-initialising of the centroids (θ_j) or increasing the number of clusters (J).
20:   **Update centroids:**
21:     Update the centroid location for each cluster as
22:     $\theta_j = \frac{1}{K_j} \sum_{k \in \mathcal{K}_j} q_k$
23: **Until** There is no change in cluster members, or the procedure reaches N_{max}.
24: **output:** $\{J^{real}\}, \{\theta_j\}$ and $\{\mathcal{K}_j = \{1, ..., K_j\}\}, \forall j$.

Instead of (14.3), we define a new expression for the flying time of a UAV between two adjacent clusters as follows:

$$\delta_{clus} = \sum_{j=1}^{J-1} \delta_{m,j}^{clus} = \sum_{j=1}^{J-1} \frac{\|\mathbf{q}_j[j+1] - \mathbf{q}_j[j]\|}{V} \tag{14.21}$$

where $\mathbf{q}_j[j] = \boldsymbol{\theta}_j$ denotes the location of the centroid of the jth cluster. Meanwhile, the initial and final flying time of the UAV over one cluster are expressed as

$$\delta_j^{st} = \frac{\|\mathbf{q}_{G,m} - \mathbf{q}_k|_{k=1,k\in\mathcal{K}_j}\|}{V} \tag{14.22}$$

$$\delta_j^{fi} = \frac{\|\mathbf{q}_{G,m} - \mathbf{q}_k|_{k=K_j}\|}{V} \tag{14.23}$$

Furthermore, the number of GSs in each cluster is limited by the maximum operating time of the UAV. The estimated flying time of the clusters over their GSs must follow the inequality

$$L_j^{clus} + \delta_j^{st} + \delta_j^{fi} \le T_{max}, \tag{14.24}$$

where

$$L_j^{clus} = \sum_{k\in\mathcal{K}_j} \left(L_{k,con}^{max} + \frac{\bar{D}_k}{\bar{C}_k} \right) \tag{14.25}$$

in order to control the number of GSs in each cluster.

14.4.2   Proposed solving approach

We assume that the UAVs communicate with the groups of GSs and gather data at the same time, during which the location of the UAVs is unchanged. To this end, binary variables $\mathbf{a}_m = [a_{m,j}]_{j=1}^J$ are defined as

$$a_{m,j} = \begin{cases} 1, & \text{UAV serves group } j \\ 0, & \text{otherwise} \end{cases} \tag{14.26}$$

Then, we propose an efficient procedure in Algorithm 16 to estimate the number of UAVs and calculate the completion time for clustering UAV-WSN under the assigning scenarios of (14.25).

Algorithm 16 Estimating the number of UAVs (M_{clus})

1: **Input:** The initial UAVs's locations ($\mathbf{q}_m[0]$) and the database of GSs clustering ($\{\boldsymbol{\theta}_j\}, \{K_j\}, J^{real}$). Set $\mathbf{a}_m = 0$
2: Set $M_{clus} = \emptyset$. Set $m = 1, J_m = \emptyset$ and $L_m^{clus} = 0$.
3: **for** $j = 1$ to J^{real} **do**
4:    **if** $L_m^{clus} + \delta_m^{st} + \delta_m^{fi} \le T_{max}$.
5:       Calculate $L_m^{clus} = L_m^{clus} + L_j^{clus} + \delta_j^{st} + \delta_j^{fi}$.
6:       Set $a_{m,j} = 1$ and add $J_m = \{j\}$.
7:    **end if**
8:    **return** J_m and $M_{clus} = \{m\}$.
9:       Set $m = m + 1, J_m = \emptyset$ and $L_m^{clus} = 0$.
10: **end for**
11: **Output:** $\{M_{clus}\}$ and $\{J_m = \{1, ..., J_m\}\}, \forall m$.

The flying time of the UAVs over their GSs for completing the collection of data packets must follow the inequality

$$L_{tot}^{clus} = \sum_{m\in\mathcal{M}_{clus}} L_m^{clus} \tag{14.27}$$

where

$$L_m^{clus} = \delta_m^{st} + \delta_m^{fi} + \sum_{j \in \mathcal{J}_m} \delta_{m,j}^{clus} + \sum_{k \in \mathcal{K}_j} \left(L_{k,con} + \frac{\bar{D}_k}{\bar{C}_k} \right),$$ (14.28)

$\delta_m^{st} = \dfrac{\|\mathbf{q}_{G,m} - \mathbf{q}_j|_{j=1, j \in \mathcal{J}_m}\|}{V}$ and $\delta_m^{fi} = \dfrac{\|\mathbf{q}_{G,m} - \mathbf{q}_j|_{j=J_m}\|}{V}$ such that

$$L_m^{clus} \leq T_{max}.$$ (14.29)

Unlike Section 14.3, this approach does not provide any optimisation problem that requires optimisation tool or optimisation procedure. Thus, this approach is presenting a simple and fast method by using Algorithms 15–16.

14.5 Simulation results

For performance evaluation, we provide a computational platform with a PC using AMD Ryzen 7 2700X, CPU @3.7 GHz and 32 GB memory.

For simulation, we set the system parameters as

- The low-power Bluetooth is used for IoT connectivity with frequency 2.4 GHz, connectivity range 100 m for maximum, data rate $\bar{C}_k = 200$ kbps and bandwidth $B = 1$ MHz)
- The data packet of all GSs is the same at $\bar{D}_k = 1$ Mbits.
- The network area is a rectangle area $\mathcal{A}(\ell, d) = (2\,000, 200)$.
- The location of the ground station and parking docks are approximated at $(100, 1\,000, 0)$.
- The number of GSs is set as $K = \{100, 200, 500\}$.
- The average flight speed and UAVs' altitude are at $V = 10$ m/s and $H_m = [20, 100]$ m.
- The operation time of the UAVs is set as $T_{max} = 20$ minutes.
- The constant transmit power of the GSs is 100 mW.
- The noise power is −90 dB.
- The connection times of K GSs ($L_{k,con}$) are randomly distributed in the interval [0.5, 5] s.

In our simulations, conventional (Conv.) path planning refers to the conventional approach by using P2P scenario without path planning optimisation. In this scheme, we set $\mathbf{q}_m[k] = \mathbf{q}_k|_{k \in \mathcal{K}_m}, \forall k$.

Figures 14.2–14.4 plot the completion time of our proposed approaches for the considered UAV-WSN versus the coverage region corresponding to the QoS requirement. In these results, the number of UAVs used was estimated for gathering data from the GSs within limited flying time. As we can see, the performance of P2P path planning scheme significantly outperforms other schemes. Interestingly, our schemes achieve the same completion time as the small coverage region. Meanwhile, P2P and CUN schemes provide better performance than Conv. scheme

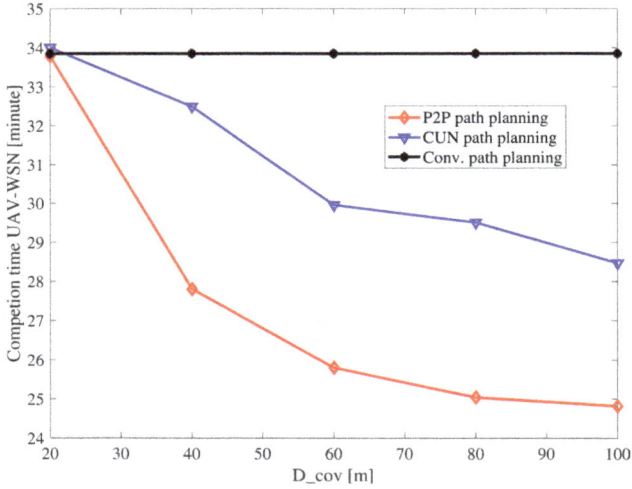

Figure 14.2   The completion time for Conv., P2P and CUN schemes versus the coverage region D_{cov} under $K = 100$. ©IEEE 2018. Reprinted with permission from [277].

when the coverage region increases. This is because the P2P and CUN approaches are more flexible and efficient for providing a short flying time in an increase of coverage region.

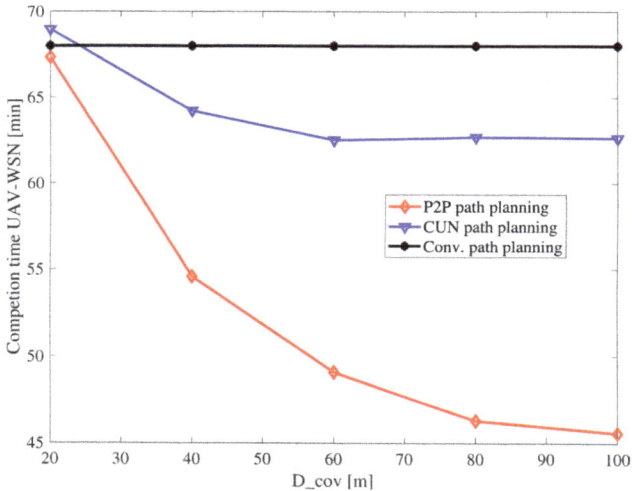

Figure 14.3   The completion time for Conv., P2P and CUN schemes versus the coverage region D_{cov} under $K = 200$. ©IEEE 2018. Reprinted with permission from [277].

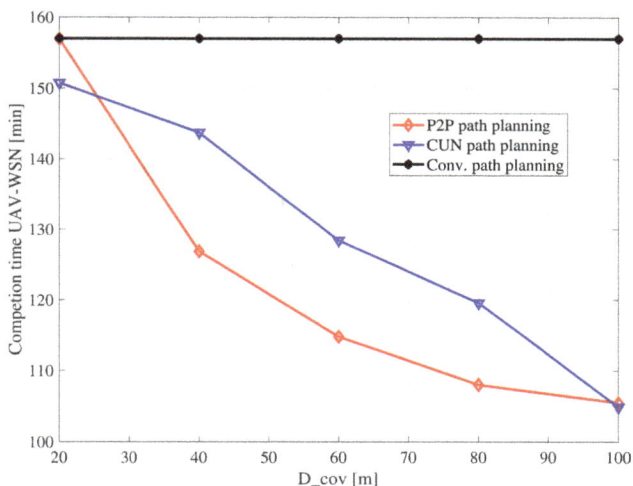

Figure 14.4   The completion time for Conv., P2P and CUN schemes versus the coverage region D_{cov} under $K = 500$. ©IEEE 2018. Reprinted with permission from [277].

Next, the computation time of Conv., P2P and CUN methods is shown in Table 14.1. By using K-means clustering model, the UAV-assisted system can save a lot of execution time compared to P2P scheme. For $K = 100$, the speed up time of CUN method is more than 300 times compared to that of the P2P. Although the CUN requires a small amount of computation time in milliseconds, the total time for the CUN to collect sensor data is significantly lower than that for P2P. Obviously, there is a trade-off between the solving time and the completion time, which should be studied in real-time applications, e.g., disaster relief.

On the other hand, the requirement of the system is also investigated by the estimated number of UAVs and the number of clusters for CUN. The good service is defined by satisfactorily tracking in the coverage region (D_{cov}) and completion of data collection. For instance, 100, 200 and 500 GSs will guarantee the QoS with the average number of UAVs used at 2, 4 and 9 for completing data collection, respectively.

Table 14.1   The executive solving times in different computational methods for solving the problem (14.13) under $D_{cov} = 100$

Number of GSs	Conv. (ms)	P2P (s)	CUN (ms)
100	3.7	4.14	12.2
200	4.3	8.22	28.7
500	5.9	26.55	137.7

14.6 Conclusions

This chapter has investigated the practical embedded optimisation methods based on the application of UAV-communicated WSN for tracking and gathering sensed data. Our low-complexity of optimisation algorithms demonstrate the effectiveness of the proposed approaches as running time for solving them can be conducted in milliseconds. We have shown that our real-time optimisation is very suitable for UAV application where the real-time control is a crucial issue. This chapter has also opened a number of future research directions in UAV for IoT wireless communication networks.

Chapter 15

Learning-aided real-time performance optimisation of cognitive UAV-assisted disaster communication

This chapter proposes efficient optimisation methods[a] for relay-assisted unmanned aerial vehicles (UAVs) in cognitive radio networks (CRNs) to cope with the network destruction in the event of a natural disaster. The model considers real-time optimisation in embedded UAV-CRN communication invoked for recovering wireless communication services. Particularly, by conceiving advanced optimisation techniques and training deep neural networks (DNNs), proposed solutions become capable of supporting real-time applications in disaster recovery scenarios. The algorithms impose low computational complexity, hence, have a low execution time in solving real-time optimisation problems. Numerical results demonstrate the benefits of our approaches proposed for UAV-CRN.

15.1 Introduction

In the event of a natural disaster, UAVs play a significant role in search and rescue (SAR) missions [97]. The UAVs have to stay airborne above the affected area to aid first responders in assessing the gravity of the disaster as promptly as possible. Yet the UAVs' airborne duration is limited by their battery capacity [97], whereas SAR missions require intensive assistance from UAVs during the first hours of the disaster.

UAVs' operation is conventionally mandated in the unlicensed spectrum bands shared with other wireless technologies including the IEEE S-Band, IEEE L-Band and ISM-Band. These bands are getting more crowded due to the escalating proliferation of Internet-of-things devices and device-to-device (D2D) communications. Hence, supporting the UAVs' operation in a CRN becomes a promising technique of increasing the UAV's available radio resources in addition to the unlicensed band. The integration of UAVs into spectrum-sharing networks has attracted substantial interest from the research community [288, 289, 290]. In [288], the authors

---

[a]This chapter has been as part of [287]

enhanced the spectrum sensing performance by arranging for a UAV to perform spectrum sensing by circularly flying over the primary user (PU) with the objective of accessing the idle spectrum. By contrast, the UAV can also operate concurrently with the PU [289], where it acts as a relay to forward the messages from both the PU and SU to the designated receivers.

Though combining a UAV with CRNs is capable of improving the spectral efficiency, there are several technical problems associated with UAV-aided communication. One of the most important issues is the UAV's energy consumption, which is the main drawback of UAV's applications [291, 292]. To address this, joint trajectory and power allocation optimisation has been conceived for UAV CRNs in [291]. Given this transmission strategy, the average achievable rate of the UAV to SU link can be optimised subject to the UAV's speed, location and transmit power. Although the aforementioned contributions have shed light on the UAVs' application, especially on their suitability in disaster-relief efforts, UAV-enabled communication is still facing limitations that should be addressed for ensuring the success of SAR missions. In particular, a prompt action is required of the network controller in support of UAV communications due to the dynamically changing environment [97], which is one of the most critical constraints in UAV applications. In all the UAV-aided optimisation scenarios found in the open literature [288, 289, 290, 291, 292] and in the references therein, solving a convex optimisation problem can only be achieved after a long period of time, which is not particularly suitable for disaster relief. Therefore, maximising the performance of UAV communication networks is vital for such applications. In fact, our recent contributions [148, 259] have demonstrated the feasibility of model the free-space path loss model, applicable with just a small amount of time needed for solving on a micro- or milli-second timescale.

Against this background, we propose a practical optimisation technique for enabling cognitive UAV communications to restore reliable network coverage in disaster-relief missions. Explicitly, joint execution time and energy efficiency (EE) optimisation is conceived, which involves the deployment of UAVs under the control of mix-integer optimisation programming and robust resource allocation under EE maximisation. The main contributions are as follows:

- An amalgamated optimisation and machine learning method relying on a DNN model is proposed for a significant reduction of the execution time under real-time solution of mixed-integer UAV deployment problems. This technique results in a learning-based optimisation programming, which we associate with the connotation of 'black box' optimisation.
- Low-complexity resource allocation is proposed for solving the nonconvex EE maximisation of our UAV deployment problem under both power budget and quality-of-service (QoS) constraints for dealing with the challenges of limited spectral and energy resources of the UAV systems.

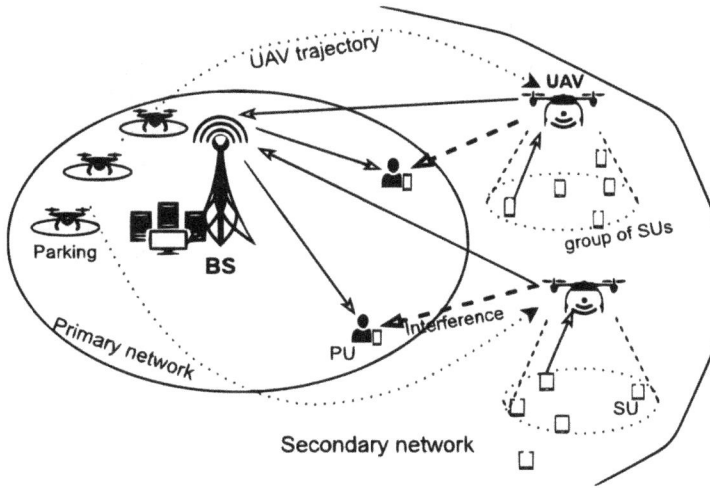

Figure 15.1   A model of UAV-enabled cognitive small cell network in disaster relief. ©IEEE 2018. Reprinted with permission from [287].

15.2   UAV-CRN system and channel model

15.2.1   System model

We consider a downlink (DL) transmission scenario, where a macro base station (MBS) is equipped with a massive multiple-input multiple-output (MIMO) array. Here, the N transmit antennas at the MBS are utilised to serve K_P PUs located in the primary network (safety area). Meanwhile, in the secondary network (disaster area), the UAVs are deployed as small-cell flying base stations (SBSs),[b] which can be connected to the cellular networks via the MBS; the aim is to restore reliable wireless network(s) operation in the hazardous areas and to serve as many SUs in the disaster region as possible. All SUs that are served are represented by M groups given by the set of $\mathcal{K}_S = \{\mathcal{K}_1, ..., \mathcal{K}_M\}$, which are supported by the set of UAVs $\mathcal{M} = \{1, ..., M\}$ required for restoring reliable network operation. We set the number of PUs and SUs to $\mathcal{K}_P = \{1, ..., K_P\}$ and $\mathcal{K}_S = \{1, ..., K_S\}$, respectively. Both the PUs and SUs are randomly distributed in the primary and secondary networks constituted by the set of $\mathcal{K} = \{\mathcal{K}_P, \mathcal{K}_S\}$. The deployment and trajectory design of UAVs is controlled by the BS on the ground, as shown in Figure 15.1. The MBS is equipped with large-scale DL MIMO antennas, while other terminals are equipped with a single antenna.

---

[b]Hereafter, BSs represent both the MBS and SBSs.

15.2.2   Channel model

Without loss of generality, we define the locations of the BS, the UAVs and all the users (PUs and SUs) as (x_0, y_0, H_0), (x_m, y_m, H_m), $m \in \mathcal{M}$ and $(x_k, y_k, 0)$, $k \in \mathcal{K}$, respectively. The antenna height of the BS and the UAV altitude are, respectively, denoted as H_0 and H_m. These locations are determined and stored at the ground station by using the global positioning system.

Due to the LOS propagation and the 3D nature of UAV-enabled communications, we can exploit the air-to-air (ATA) link to enhance the links BS–BS explicitly, due to the fact that LoS propagation is highly likely to occur in the ATA links, the path loss of BS–BS (BS–UAV and UAV–UAV) follows the free-space path loss model as in [239, 259].

$$\beta_{m,n}^{ata} = \beta_0 R_{m,n}^{-2}, \ m, n = 0, 1, ..., M \tag{15.1}$$

where β_0 is the channel's power gain at reference distance d_0 and $R_{m,n}$ denotes the distance between the mth BS and the nth BS which is given by

$$R_{m,n} = \sqrt{d_{m,n}^2 + (H_m - H_n)^2} \tag{15.2}$$

where $d_{m,n} = \sqrt{(x_m - x_n)^2 + (y_m - y_n)^2}$.

By contrast, the air-to-ground (ATG) channels are more complex due to the effects of propagation blockage of shadowing, blockage geometry and disaster paraphernalia. The path-loss expression between the mth BS and the kth user is denoted as [241]

$$\beta_{m,k}^{atg} = PL_{m,k} + \eta_{LoS} P_{m,k}^{LoS} + \eta_{NLoS} P_{m,k}^{NLoS} \tag{15.3}$$

where η_{LoS} and η_{NLoS} are the average additional losses for the LoS and NLoS paths, respectively. Here, the distance-related path loss is given by

$$PL_{m,k} = 10 \log \left(\frac{4\pi f_c R_{m,k}}{c} \right)^{\alpha} \tag{15.4}$$

where f_c is the carrier frequency (Hz), c is the speed of light (m/s) and $\alpha \geq 2$ is the path loss exponent. The probability of LoS is given by [242]

$$P_{m,k}^{LoS} = \frac{1}{1 + a \exp \left[-b \left(\arctan \left(\frac{H_{U,m}}{d_{m,k}} \right) - a \right) \right]} \tag{15.5}$$

where a and b are constants, depending on the environment. Then, we have $P_{m,k}^{NLoS} = 1 - P_{m,k}^{LoS}$. Finally, we can rewrite (15.3) as

$$\beta_{m,k}^{atg} = 10\alpha \log(R_{m,k}) + A \times P_{m,k}^{LoS} + B \tag{15.6}$$

where $A = \eta_{LoS} - \eta_{NLoS}$, $B = PL_{m,k} + \eta_{NLoS}$ and $R_{m,k}$ denotes the distance between BS m and user k, formulated as

$$R_{m,k} = \sqrt{d_{m,k}^2 + H_m^2}, k \in \mathcal{K} \tag{15.7}$$

where $d_{m,k} = \sqrt{(x_m - x_k)^2 + (y_m - y_k)^2}$ is the Euclidean distance between UAV m and user k.

On the other hand, the small-scale fading of all channels ($\mathbf{h}_{m,0}^{ata}$, $\mathbf{h}_{0,k}^{atg} \in \mathbb{C}^N$ and $h_{m,n}^{ata}$, $h_{m,k}^{atg} \in \mathbb{C}$) is assumed to obey an i.i.d. random variable with zero mean and unit variance. Hence, the full ATA and ATG channel models can be exploited as $g_{m,n} = \sqrt{\beta_{m,n}^{ata}} h_{m,n}^{ata}$ and $g_{m,k} = \sqrt{\beta_{m,k}^{atg}} h_{m,k}^{atg}$, respectively.

15.2.3 Transmission scheme

Let us consider the DL transmission phases where the MBS transmits its signal to the PUs. First, the signal received at the PUs is given by

$$y_{0,k} = \underbrace{\sqrt{P_0}\mathbf{g}_{0,k}^T \mathbf{f}_{0,k} s_{0,k}}_{\text{desired signal}} + \underbrace{\sum_{k' \in \mathcal{K}_p \backslash \{k\}} \sqrt{P_0} \mathbf{g}_{0,k}^T \mathbf{f}_{0,k'} s_{0,k'}}_{\text{co-tier interference}} + \underbrace{\sum_{l=1}^{M} g_{l,k}^{atg} \sqrt{P_l} s_{l,0}}_{\text{inter-cell interference}} + n_k \tag{15.8}$$

where P_0 is the transmit power of the BS; $\mathbf{f}_{0,k} \in \mathbb{C}^N$ and $s_{0,k} \in \mathbb{C}$ are the beamforming vector and the information transmitted from the MBS with $\|s_{0,k}\|^2 \leq 1$ is the additive white Gaussian noise (AWGN) and P_l is the transmit power of the lth UAV.

In this chapter, we employ efficient maximal ratio transmission (MRT) criterion-based beamforming design for our system, which is formulated as:

$$\mathbf{f}_{0,k} = \sqrt{p_{0,k}} \frac{\mathbf{g}_{0,k}^*}{\|\mathbf{g}_{0,k}\|} \tag{15.9}$$

where $p_{0,k}$ is the power control coefficient. Then, we introduce $\rho_{0,k,j} = \mathbf{g}_{0,k}^T \mathbf{g}_{0,j}^* / \|\mathbf{g}_{0,j}\|$.

For $\mathbf{p}_0 = [p_{0,k}]_{k \in \mathcal{K}_P}$ and $\mathbf{p}_M = [P_m]_{m \in \mathcal{M}}$, the interference imposed on the primary network is characterised by the co-tier interference formulated as

$$\mathcal{I}_k^{\text{intra}}(\mathbf{p}_0) = P_0 \sum_{k' \in \mathcal{K}_P \backslash \{k\}} p_{0,k'} |\rho_{0,k,k'}|^2, k \in \mathcal{K}_P \tag{15.10}$$

and the inter-cell interference inflicted by the secondary network[c] is

$$\mathcal{I}_k^{\text{inter}}(\mathbf{p}_M) = \sum_{m \in \mathcal{M}} P_m |\beta_{m,k}^{atg}|^2, k \in \mathcal{K}_P. \tag{15.11}$$

The information throughput of the kth PU (in nats) is

$$R_{0,k}(\mathbf{p}_0, \mathbf{p}_M) = \ln\left(1 + \frac{P_0 p_{0,k} |\rho_{0,k,k}|^2}{\mathcal{I}_k^{\text{intra}}(\mathbf{p}_0) + \sigma_k^2}\right). \tag{15.12}$$

---

[c]It is very hard to estimate the ATG channel between UAVs and PUs, so the inter-cell interference from secondary networks can only be estimated by the UAVs and determined as in (15.11).

To ensure the QoS of the primary network, the QoS constraints have to be investigated in the face of inter-cell interference

$$\mathcal{I}_k^{\text{inter}}(\mathbf{p}_M) \leq I_{th}^{PU} \tag{15.13}$$

where I_{th}^{PU} is the maximum tolerable interference still capable of ensuring the QoS of the PUs.

Simultaneously, the UAVs also transmit the signal to the MBS, which will then forward the signal from the SUs to the MBS. The signal received at the MBS from the UAV m is written as

$$y_{m,0} = \underbrace{\mathbf{g}_{m,0}^T \mathbf{f}_{m,0} \sqrt{P_m} s_{m,0}}_{\text{desired signal}} + \underbrace{\sum_{l=1,l\neq m}^{M} \mathbf{g}_{m,0}^T \mathbf{f}_{l,0} \sqrt{P_l} s_{l,0}}_{\text{inter-cell interference}} + n_0 \tag{15.14}$$

where $\mathbf{f}_{m,0}$ is transmit beamforming vector and $s_{m,0}$ is information transmitted by UAV m with $\|s_{m,0}\|^2 \leq 1$, $n_0 \sim \mathcal{CN}(0, \sigma_0^2)$ is the AWGN.

Similar to (15.9), we apply MRT for this case and we also introduce $\rho_{m,0,l} = \mathbf{g}_{m,0}^T \mathbf{g}_{l,0}^*/\|\mathbf{g}_{l,0}\|$.

The information throughput of the MBS (in nats) may be written as

$$R_{m,0}(\mathbf{p}_M) = \ln\left(1 + \frac{P_m|\rho_{m,0,m}|^2}{\mathcal{I}_m^{\text{MBS}}(\mathbf{p}_M) + \sigma_0^2}\right) \tag{15.15}$$

where $\mathcal{I}_m^{\text{MBS}}(\mathbf{p}_M) = \sum_{l\in\mathcal{M},l\neq m} P_l|\rho_{m,0,l}|^2$ represents the inter-cell interference imposed on the MBS.

Thus, the total throughput of the DL phase considered is expressed as

$$R_{tot}(\mathbf{p}_0, \mathbf{p}_M) = \sum_{k\in\mathcal{K}_P} R_{0,k}(\mathbf{p}_0, \mathbf{p}_M) + \sum_{m\in\mathcal{M}} R_{m,0}(\mathbf{p}_M). \tag{15.16}$$

15.2.4   Problem formulation

In this treatise, our target is to maximise the network's EE using BS association and power allocation optimisation for cognitive UAV communication. The corresponding EE maximisation problem is formulated as

$$\max_{\mathbf{p}_0,\mathbf{p}_M,(m,k)} \frac{B \cdot R_{tot}(\mathbf{p}_0, \mathbf{p}_M)}{\text{Pow}_{tot}} \tag{15.17a}$$

$$s.t \quad (15.13) \tag{15.17b}$$

$$\sum_{k\in\mathcal{K}_P} p_{0,k} \leq 1, P_m \leq P_m^{max}, m \in \mathcal{M} \tag{15.17c}$$

$$R_{m,0}(\mathbf{p}_M) \geq \bar{r}_{m,0}, \ m \in \mathcal{M}$$

$$\sum_{k\in\mathcal{K}_P} p_{0,k} \leq 1, P_m \leq P_m^{max}, m \in \mathcal{M} \tag{15.17d}$$

$$R_{0,k}(\mathbf{p}_0, \mathbf{p}_M) \geq \bar{r}_{0,k},\ k \in \mathcal{K}_P \tag{15.17e}$$

$$(m, k) \in \mathcal{K}_m, m \in \mathcal{M},\ k \in \mathcal{K}_m \tag{15.17f}$$

where B is the bandwidth of the system. The constraint (15.17c) represents the power requirement at the MBS and the UAVs, whereas constraints (15.17d) and (15.17e) formulate the QoS requirement of the MBS-UAV and MBS-PU links. The constraint (15.17f) corresponds to the deployment of the UAVs at the beginning. We set $\mathcal{K}_m = \{1, ..., K_m\}$ and $\sum_{m\in\mathcal{M}} K_m = K_S$.

The total power consumption in the transmission phase considered is defined as

$$\text{Pow}_{tot} = \eta_0 P_0 \sum_{k\in\mathcal{K}_P} p_{0,k} + P_{c,0} + \sum_{m\in\mathcal{M}}(\eta_m P_m + P_{c,m}) \tag{15.18}$$

where $\eta_0, \eta_m > 0$ are the reciprocal drain efficiency of the power amplifier at the MBS and UAVs; $P_{c,0}$ and $P_{c,m}$ denote the power consumption of signal processing at the MBS and the UAVs, respectively.

For efficiently solving the nonconvex problem (15.17), we separate the problem (15.17) into two subproblems. First, the user association with UAV clustering will be proposed that will satisfy the constraint (15.17e) under the deployment of UAVs. Then, a DNN is applied for constructing the optimisation strategy of UAV deployment for the real-time context considered. Finally, the optimal power is assigned for maximising the network's EE in the face of constrained power budget and QoS requirement.

15.3 Learning optimisation for a real-time scenario of UAV deployment

15.3.1 *Conventional optimisation approach for UAV deployment*

In order to guarantee the QoS of ATG links between UAVs and users, we consider the coverage region by defining a circular disc of radius D_{cov}. The radius $D_{m,cov}$ is related to the altitude of the UAV m as follows

$$H_m = D_{m,cov} \tan(\theta), \forall m \tag{15.19}$$

where θ is set to $42.44°$ [281]. Therefore, a SU can be served by a UAV in its coverage area $(m, k) \in \mathcal{K}_m$ if the Euclidean distance between the UAV and the SU is lower than the coverage distance $D_{m,cov}$, which is formulated as

$$d_{m,k} \leq D_{m,cov}^{max},\ k \in \mathcal{K}_S \tag{15.20}$$

where $D_{m,cov}^{max} = H_m^{max}/\tan(\theta)$.

Given the limited operational range of the UAV, we formulate a UAV positioning optimisation problem to provide a best-effort transmission service for the secondary network in each group as follows:

$$\max_{q_m, u_{m,k}} \sum_{m=1}^{M} \sum_{k=1}^{K_m} u_{m,k} \tag{15.21a}$$

s.t. $d_{m,k}^2 \leq (D_{m,cov}^{max})^2 + \lambda_m(1 - u_{m,k})$ $\qquad$ (15.21b)

$\mathbf{q}_m \in [\mathbf{q}_m^{min}, \mathbf{q}_m^{max}]$ $\qquad$ (15.21c)

$u_{m,k} \in \{0, 1\}, \; m \in \mathcal{M}, \; k \in \mathcal{K}_m$ $\qquad$ (15.21d)

where $\mathbf{q}_m = [x_m, y_m, H_m^{max}]^T$, λ_m is chosen as a specific value corresponding to the maximum network coverage area of the mth UAV (i.e., $\lambda_m > (D_{m,cov}^{max})^2$), while (x_{min}, x_{max}) and (y_{min}, y_{max}) represent the lower and upper bounds of the horizontal and vertical range of UAVs, respectively. Note that the problem in (15.21) is a mixed-integer (binary) quadratic programming, which is a nonconvex problem. Fortunately, the Python-embedded optimisation program CVXPY [223] is efficiently capable of solving problem (15.21).

15.3.2   Deep neural network for learning optimisation of UAV deployment

Although conventional optimisation relying on the CVXPY package, for example, can solve problem (15.21), the execution time imposed by solving the related mixed-integer program is excessive, when the networking scenario becomes more complex and when the number of integer variables ($u_{m,k}$) increases.

In this work, to tackle the aforementioned problem, we apply a new optimisation technique eminently suitable for real-time applications by amalgamating DNN and optimisation algorithms. This technique results in learning-based optimisation programming [293] as presented in Figure 15.2. A DNN model is with three types of layers including an input layer, multiple hidden layers and an output layer.

In this context, a learning-based approach is proposed for the optimisation of wireless networks in [293]. Explicitly, the optimisation algorithms will be trained for learning the input/output relationship by using a DNN model during the training stage. With the aid of sufficient training data set, their optimisation technique is capable of completely replacing the conventional optimisation processes during the testing stage.

Following the system setup in [293], we configure the network structure for our DNN model as follows:

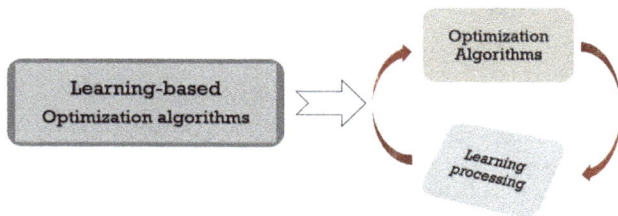

Figure 15.2   A model of learning-based optimisation algorithm by using DNN. ©IEEE 2018. Reprinted with permission from [287].

- The input of the network is the location of the UAVs and SUs $(\mathbf{q}_m, \mathbf{q}_k^{SU})$, whereas the output of the network is the optimal value of $\mathbf{q}_m^*$. In all layers, we use 'sigmoid' as the activation function.
- In the testing stage, we use a large training data set $(\mathbf{q}_m, \mathbf{q}_k^{SU})$ for optimising and learning the weights of the DNN model. The cost function is the mean squared error and the mini-batch stochastic gradient descent optimisation algorithm is used [293].
- In the testing stage, we also generate the structure based on the same distribution during the training stage. Each distributed location experiment is passed through the trained network and then we collect the resultant optimal location of the UAVs.

15.4 Maximising EE performance via robust power allocation

Having discussed the UAV deployment philosophy, let us now conceive efficient resource allocation for solving the EE maximisation problem (15.17) in the absence of nonconvex user association constraints $(u_{m,k})$. On the other hand, problem (15.17) is still a nonconvex one, since the objective function is nonconcave. Hence, we consider the modified problem of

$$\max_{\mathbf{p}_0, \mathbf{p}_M} \frac{B \cdot R_{tot}(\mathbf{p}_0, \mathbf{p}_M)}{\text{Pow}_{tot}} \tag{15.22a}$$

$$\text{s.t.} \quad (15.13), (15.17c), (15.17d). \tag{15.22b}$$

To solve problem (15.22), we exploit the logarithmic inequality of [148, 294], which follows from the convexity of the function $f(x, y) =$, yielding

$$f(x, y) = \ln(1 + \frac{1}{xy}) \geq \hat{f}(x, y) \tag{15.23}$$

where we have

$$\hat{f}(x, y) = \ln\left(1 + \frac{1}{\bar{x}\bar{y}}\right) + \frac{2}{(\bar{x}\bar{y} + 1)} - \frac{x}{\bar{x}(\bar{x}\bar{y} + 1)} - \frac{y}{\bar{y}(\bar{x}\bar{y} + 1)} \tag{15.24}$$

$$\forall x > 0, \bar{x} > 0, y > 0, \bar{y} > 0$$

Let i denote the ith iteration and exploit

$$x = \frac{1}{P_0 p_{0,k} |\rho_{0,k,k}|^2}, \quad y = I_k^{\text{intra}}(\mathbf{p}_0) + \sigma_k^2,$$

$$\bar{x} = x^{(i)} = \frac{1}{P_0 p_{0,k}^{(i)} |\rho_{0,k,k}|^2}, \quad \bar{y} = y^{(i)} = I_k^{\text{intra}}(\mathbf{p}_0^{(i)}) + \sigma_k^2$$

for the approximation of the kth PU's throughput in (15.12) is

$$R_{0,k}(\mathbf{p}_0, \mathbf{p}_M) \geq \hat{R}_{0,k}^{(i)}(\mathbf{p}_0, \mathbf{p}_M), \forall k \in \mathcal{K}_P \tag{15.25}$$

Similarly, we can invoke

$$x = \frac{1}{P_m |\rho_{m,0,m}|^2}, y = I_m^{MBS}(\mathbf{p}_M) + \sigma_0^2,$$

$$\bar{x} = x^{(i)} = \frac{1}{P_m^{(i)}|\rho_{m,0,m}|^2}, \bar{y} = y^{(i)} = \mathcal{I}_m^{MBS}(\mathbf{p}_M^{(i)}) + \sigma_0^2$$

for the approximation of MBS's throughput function in (15.15) as

$$R_{m,0}(\mathbf{p}_M) \geq \hat{R}_{m,o}^{(i)}(pM), m \in \mathcal{M} \tag{15.26}$$

where the form of $\hat{R}_{m,o}^{(i)}(\mathbf{p}_M)$ and $\hat{R}_{0,k}^{(i)}(\mathbf{p}_0, \mathbf{p}_M)$ is defined by (15.24). Given the feasible points $(\mathbf{p}_0^{(i)}, \mathbf{p}_M^{(i)})$ of (15.22), we arrive at

$$\phi^{(i)} = B \cdot R_{tot}(\mathbf{p}_0^{(i)}, \mathbf{p}_M^{(i)})/\text{Pow}_{tot}(\mathbf{p}_0^{(i)}, \mathbf{p}_M^{(i)}). \tag{15.27}$$

At the ith iteration, the following convex program is solved to generate the next feasible point:

$$\max_{\mathbf{p}_0, \mathbf{p}_M} B \cdot \hat{R}_{tot}^{(i)}(\mathbf{p}_0, \mathbf{p}_M) - \phi^{(i)}\text{Pow}_{tot}(\mathbf{p}_0, \mathbf{p}_M) \tag{15.28a}$$

$$\text{s.t. } (15.13), (15.19), (15.20) \tag{15.28b}$$

where we have $\hat{R}_{tot}^{(i)}(\mathbf{p}_0, \mathbf{p}_M) = \sum_{k \in \mathcal{K}_P} \hat{R}_{0,k}^{(i)}(\mathbf{p}_0, \mathbf{p}_M) + \sum_{m \in \mathcal{M}} \hat{R}_{m,0}^{(i)}(\mathbf{p}_M)$.

We now proceed by proposing an algorithm to solve the EE maximisation (15.28). The initial point $(\mathbf{p}_0^{(0)}, \mathbf{p}_M^{(0)})$ for (15.28) may be found by random search for satisfying constraint (15.28b).

Algorithm 17 : Power-allocation procedure for solving problem (15.22)

Initialisation: Set feasible points $(\mathbf{p}_0^{(0)}, \mathbf{p}_M^{(0)})$, $i = 0$ and $\phi^{(0)} = B \cdot R_{tot}(\mathbf{p}_0^{(0)}, \mathbf{p}_M^{(0)})/\text{Pow}_{tot}(\mathbf{p}_0^{(0)}, \mathbf{p}_M^{(0)})$. Set the tolerance $\varepsilon = 10^{-3}$ and the maximum number of iterations $I_{max} = 20$.

Repeat

  Solve the (15.28) for the optimal solution $(\mathbf{p}_0^{(i+1)}, \mathbf{p}_M^{(i+1)})$. Set $\phi^{(i+1)} = B \cdot R_{tot}(\mathbf{p}_0^{(i+1)}, \mathbf{p}_M^{(i+1)})/\text{Pow}_{tot}(\mathbf{p}_0^{(i+1)}, \mathbf{p}_M^{(i+1)})$.
Set $i := i + 1$

Stop convergence of the objective in (15.28) or $i > I_{max}$.

Note that the optimal solution of $\{\phi^*\}$ in (15.27) will be an optimal parameter maximising the EE formulated in (15.22) for the entire network.

15.5   Simulation results

In this section, the performance of the network considered is evaluated by using embedded optimisation programming, such as, for example, the CVXPY version 1.0.21 in Python [223]. The computational platform includes a PC having a AMD

Ryzen 7 2700X, CPU @3.7 GHz and 32 GB memory. Our DNN model is implemented in Python 3.6 associated with Keras 2.2.4 using TensorFlow 1.13.1.

We set the system parameters for our simulations as follows:

- The safety area is a circle coverage with a radius of 500 m.
- The disaster area is extended from the safety area with a radius up to 2 000 m.
- The location of the BS is assumed at $(0, 0, 30)$ whereas the locations of PUs and SUs are randomly distributed in the primary network and secondary network, respectively.
- The path loss from MBS to PUs is given as $\beta_{0,k}^{atg} = 148.1 + 37.6 \log_{10} R$ [dB], where R is in km.
- The number of UAVs is provided as $M = \{2, 4, 8\}$. The number of PUs is set to $K_P = \{5, 10, 20\}$ whereas the number of SUs in each group is set to $K_m = \{10, 20, 30\}$
- The limited altitude of the UAVs (H^{min}, H^{max}) is $(50, 150)$ m.
- The tolerance and the maximum number of iterations for convergence of algorithms are $\varepsilon = 10^{-3}$ and $I_{max} = 20$.
- The carrier frequency/bandwidth is $f_c = 2$ GHz/$B = 10$ MHz.
- The QoS thresholds are set to $\bar{r}_{m,0} = 50$ Mbps and $\bar{r}_{0,k} = 1$ Mbps.
- The maximum BSs transmit power is set to 40 W and 5 W for MBS and UAVs, respectively.
- The white power spectral density is $\sigma^2 = -130$ dBm/Hz.

The tolerance and the maximum number of iterations for convergence of algorithms are $\varepsilon = 10^{-3}$ and $I_{max} = 20$. The parameters of the channel model are set as in [239, 281, 148].

Figure 15.3 portrays the convergence of Algorithm 15.4. It is observed that after a few iterations, the target function (15.22) converges to its maximum value. On the other hand, the EE performance of the system is degraded when the number of PUs increases.

Next, we show the execution time of the UAV deployment algorithm in Table 15.1. The accuracy metric is defined by the average error value between the results of UAV deployment for conventional optimisation (Conv. OPT) and DNN optimisation. The figures demonstrate the potential benefit of our optimisation technique using the DNN model. As seen from Table 15.1, our proposed UAV deployment algorithm exhibits a low complexity, even upon dealing with large-scale scenarios.

Figure 15.4 plots the EE performance of our proposed techniques versus the number of PUs. In these simulations, EE_max represents the power allocation of the proposed Algorithm 15.4, while Sumrate_max and Power_min represent the power allocation invoked for the sum-rate maximisation and power-minimisation problem by appropriately adjusting problem (15.22). As expected, EE_max provides a significant EE performance improvement over the other methods. This is because the EE_max-based power allocation can efficiently control the power of both the MBS

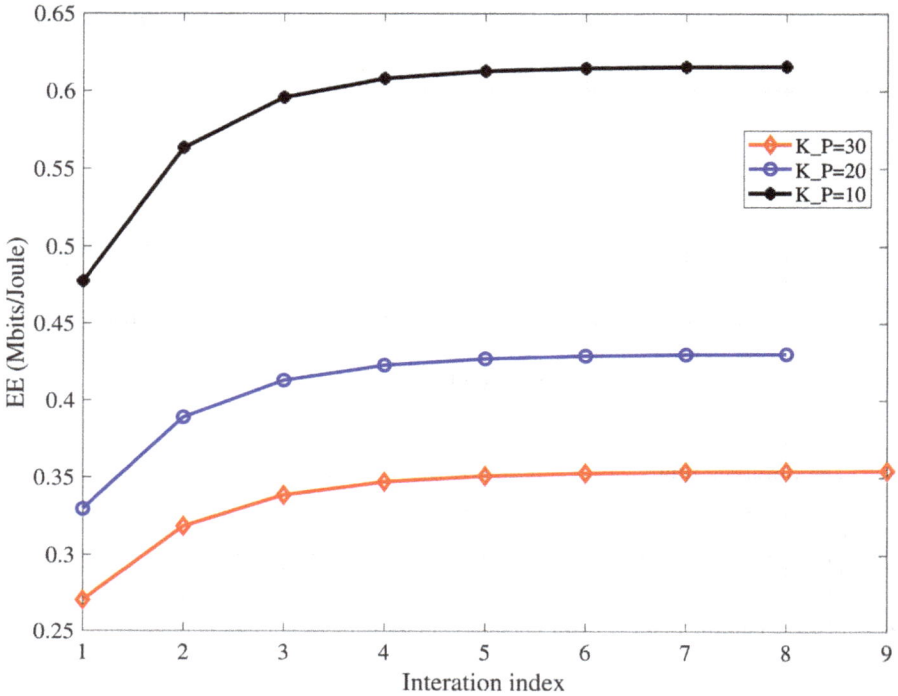

Figure 15.3   The convergence of Algorithm 15.4 under M = 4 and K_m = 20.
©IEEE 2018. Reprinted with permission from [287].

and the UAVs, while the sum-throughput can be improved despite maintaining a high QoS. On the other hand, the networks EE will be reduced upon increasing the number of PUs.

15.6   Conclusions

A novel learning-aided optimisation scheme was conceived for establishing network coverage in the event of a natural disaster. By employing the deep learning

Table 15.1   The execution time of our UAV deployment algorithm both under
Conv. OPT and learning-aided optimisation using DNN (DNN OPT)

Scenarios M, K_p, K_m	Conv. OPT (s)	DNN OPT (s)	Accuracy (%)
(2,5,10}	0.15	0.027	93.02
{4,10,20}	0.80	0.028	92.37
{8,20,30}	4.27	0.028	90.82

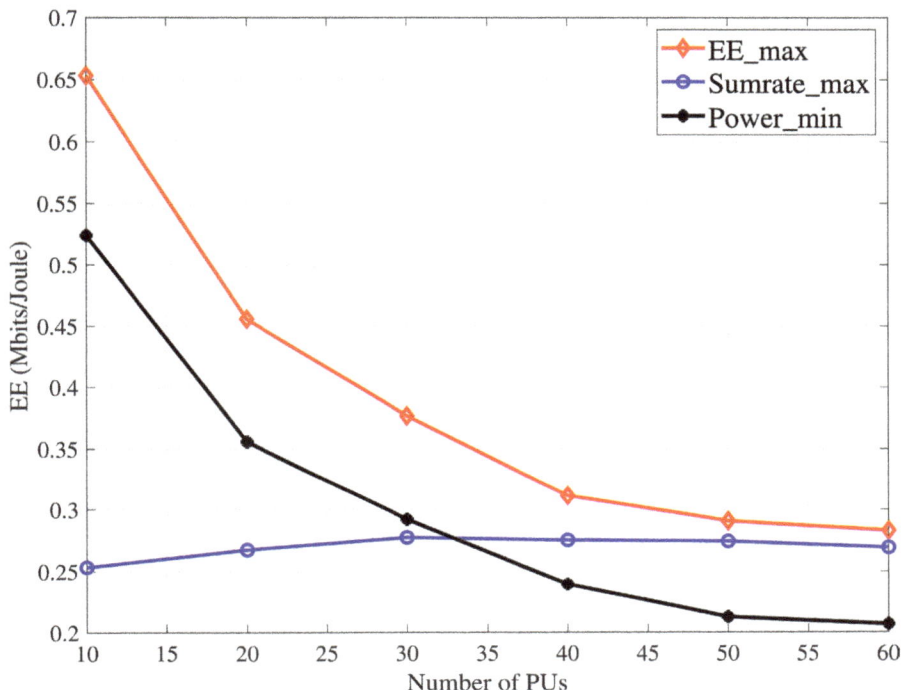

Figure 15.4 *The EE performance of proposed methods versus the number of*
PUs under M = 4 and K_m = 20. ©IEEE 2018. Reprinted with
permission from [287].

approach, our low-complexity algorithms lend themselves to real-time deployment in the context of CRNs replying on UAVs. The numerical results demonstrate that our UAV deployment can be promptly optimised in a large-scale scenario. The proposed scheme demonstrates a compelling installation of real-time optimisation in wireless communication systems destroyed by natural disasters.

References

[1] Boyd S., Vandenberghe L. *Convex Optimization*. Cambridge University Press; 2004.

[2] Zhi-Quan Luo., Wei Yu. 'An introduction to convex optimization for communications and signal processing'. *IEEE Journal on Selected Areas in Communications*. 2006, vol. 24(8), pp. 1426–38.

[3] Bengtsson M., Ottersten B. 'Optimal and suboptimal transmit beamforming'. Handbook of Antennas in Wireless Communications. CRC Press; 2001. pp. 18-1–18-33.

[4] Bengtsson M., Ottersten B. 'Optimal and suboptimal transmit beamforming'. *Handbook of Antennas in Wireless Communications, CRC Press*. 2001, vol. 01.

[5] Wiesel A., Eldar Y.C., Shamai S. 'Linear precoding via conic optimization for fixed MIMO receivers'. *IEEE Transactions on Signal Processing*. 2006, vol. 54(1), pp. 161–76.

[6] Vishwanath S., Jindal N., Goldsmith A. 'Duality, achievable rates, and sum-rate capacity of Gaussian MIMO broadcast channels'. *IEEE Transactions on Information Theory*. 2003, vol. 49(10), pp. 2658–68.

[7] Wei Yu., Lui R. 'Dual methods for nonconvex spectrum optimization of multicarrier systems'. *IEEE Transactions on Communications*. 2006, vol. 54(7), pp. 1310–22.

[8] Luo Z.-Q., Yu W. 'An introduction to convex optimization for communications and signal processing'. *IEEE Journal on Selected Areas in Communications*. 2006, vol. 24(8), pp. 1426–38.

[9] Antoniou A., W.-S. L. *Practical optimization: algorithms and engineering applications*. Springer Science, Business Media; 2007.

[10] Vandenberghe L., Boyd S. 'Semidefinite programming'. *SIAM Review*. 1996, vol. 38(1), pp. 49–95.

[11] Vorobyov S.A., Cui S., Eldar Y.C., Ma W.-K., Utschick W. 'Optimization techniques in wireless communications'. *EURASIP Journal on Wireless Communications and Networking*. 2009, vol. 2009(1), p. 567416.

[12] Tang J., So D.K.C., Alsusa E., Hamdi K.A., Shojaeifard A. 'Resource allocation for energy efficiency optimization in heterogeneous networks'. *IEEE Journal on Selected Areas in Communications*. 2015, vol. 33(10), pp. 2104–17.

[13] Zukang Shen., Andrews J.G., Evans B.L. 'Adaptive resource allocation in multiuser OFDM systems with proportional rate constraints'. *IEEE Transactions on Wireless Communications*. 2005, vol. 4(6), pp. 2726–37.

[14] Yu C.H., Doppler K., Ribeiro C.B., Tirkkonen O. 'Resource sharing optimization for device-to-device communication underlaying cellular networks'. *IEEE Transactions on Wireless Communications*. 2011, vol. 10(8), pp. 2752–63.

[15] T. C. yN., Yu W. 'Joint optimization of relay strategies and resource allocations in cooperative cellular networks'. *IEEE Journal on Selected Areas in Communications*. 2007, vol. 25(2), pp. 328–39.

[16] Sheng Z., Tuan H.D., Duong T.Q., Poor H.V. 'Joint power allocation and beamforming for energy-efficient two-way multi-relay communications'. *IEEE Transactions on Wireless Communications*. 2017, vol. 16(10), pp. 6660–71.

[17] Venturino L., Zappone A., Risi C., Buzzi S. 'Energy-efficient scheduling and power allocation in downlink ofdma networks with base station coordination'. *IEEE Transactions on Wireless Communications*. 2015, vol. 14(1), pp. 1–14.

[18] Ling J., Chizhik D., Chen C.S., Valenzuela R.A. 'Capacity growth of heterogeneous cellular networks'. *Bell Labs Technical Journal*. 2013, vol. 18(1), pp. 27–40.

[19] Caire G., Shamai S. 'On the achievable throughput of a multiantenna Gaussian broadcast channel'. *IEEE Transactions on Information Theory*. 2003, vol. 49(7), pp. 1691–706.

[20] Christensen S.S., Agarwal R., De Carvalho E., Cioffi J.M. 'Weighted sum-rate maximization using weighted MMSE for MIMO-BC beamforming design'. *IEEE Transactions on Wireless Communications*. 2008, vol. 7(12), pp. 4792–9.

[21] Yoo T., Goldsmith A. 'On the optimality of multiantenna broadcast scheduling using zero-forcing beamforming'. *IEEE Journal on Selected Areas in Communications : A Publication of the IEEE Communications Society*. 2006, vol. 54(3), pp. 528–41.

[22] Gesbert D., Hanly S., Huang H., Shamai Shitz S., Simeone O., Yu W. 'Multicell MIMO cooperative networks: a new look at interference'. *IEEE Journal on Selected Areas in Communications*. 2010, vol. 28(9), pp. 1380–408.

[23] Lopez-Perez D., Guvenc I., de la Roche G., Kountouris M., Quek T., Zhang J. 'Enhanced intercell interference coordination challenges in heterogeneous networks'. *IEEE Wireless Communications*. 2011, vol. 18(3), pp. 22–30.

[24] Shin W., Noh W., Jang K., Choi H.-H. 'Hierarchical interference alignment for downlink heterogeneous networks'. *IEEE Transactions on Wireless Communications*. 2012, vol. 11(12), pp. 4549–59.

[25] Nguyen L.D., Tuan H.D., Duong T.Q. 'Energy-efficient signalling in QoS constrained heterogeneous networks'. *IEEE Access*. 2016, vol. 4, pp. 7958–66.

[26] Nguyen D., Tran L.-N., Pirinen P., Latva-aho M. 'Precoding for full duplex multiuser MIMO systems: spectral and energy efficiency maximization'. *IEEE Transactions on Signal Processing*. 2013, vol. 61(16), pp. 4038–50.

[27] Li J., Bjornson E., Svensson T., Eriksson T., Debbah M. 'Joint precoding and load balancing optimization for energy-efficient heterogeneous networks'. *IEEE Transactions on Wireless Communications*. 2015, vol. 14(10), pp. 5810–22.

[28] Andrews J.G., Zhang X., Durgin G.D., Gupta A.K. 'Are we approaching the fundamental limits of wireless network densification?'. *IEEE Communications Magazine*. 2016, vol. 54(10), pp. 184–90.

[29] Li Y., Sheng M., Sun Y., Shi Y. 'Joint optimization of Bs operation, user association, subcarrier assignment, and power allocation for energy-efficient HetNets'. *IEEE Journal on Selected Areas in Communications*. 2016, vol. 34(12), pp. 3339–53.

[30] Nguyen V.-D., Tuan H.D., Duong T.Q., Shin O.-S., Poor H.V. 'Joint fractional time allocation and beamforming for downlink multiuser miso systems'. *IEEE Communications Letters*. 2017, vol. 21(12), pp. 2650–3.

[31] Pantisano F., Bennis M., Saad W., Debbah M., Latva-aho M. 'Improving macrocell-small cell coexistence through adaptive interference draining'. *IEEE Transactions on Wireless Communications*. 2014, vol. 13(2), pp. 942–55.

[32] Adhikary A., Dhillon H.S., Caire G. 'Massive-MIMO meets HetNet: interference coordination through spatial blanking'. *IEEE Journal on Selected Areas in Communications*. 2015, vol. 33(6), pp. 1171–86.

[33] Bhushan N., Junyi Li., Malladi D., *et al.* 'Network densification: the dominant theme for wireless evolution into 5G'. *IEEE Communications Magazine*. 2014, vol. 52(2), pp. 82–9.

[34] Hong M., Sun R., Baligh H., Luo Z.-Q. 'Joint base station clustering and beamformer design for partial coordinated transmission in heterogeneous networks'. *IEEE Journal on Selected Areas in Communications*. 2013, vol. 31(2), pp. 226–40.

[35] Abdelnasser A., Hossain E., Kim D.I. 'Clustering and resource allocation for dense femtocells in a two-tier cellular OFDMA network'. *IEEE Transactions on Wireless Communications*. 2014, vol. 13(3), pp. 1628–41.

[36] Cho H.H., Lai C.F., Shih T.K., Chao H.C. 'Integration of SDR and SDN for 5G'. *IEEE Access: Practical Innovations, Open Solutions*. 2014, vol. 2, pp. 1196–204.

[37] Andrews J.G., Buzzi S., Choi W., *et al.* 'What will 5G be?' *IEEE Journal on Selected Areas in Communications*. 2014, vol. 32(6), pp. 1065–82.

[38] Li Q.C., Niu H., Papathanassiou A.T., Wu G. '5G network capacity: key elements and technologies'. *IEEE Vehicular Technology Magazine*. 2014, vol. 9(1), pp. 71–8.

[39] Ghosh A., Mangalvedhe N., Ratasuk R., *et al.* 'Heterogeneous cellular networks: from theory to practice'. *IEEE Communications Magazine*. 2012, vol. 50(6), pp. 54–64.

[40] Mugen Peng., Yuan Li., Jiamo Jiang., Jian Li., Chonggang Wang. 'Heterogeneous cloud radio access networks: a new perspective for enhancing spectral and energy efficiencies'. *IEEE Wireless Communications*. 2014, vol. 21(6), pp. 126–35.

[41] Dhillon H.S., Kountouris M., Andrews J.G. 'Downlink MIMO HetNets: modeling, ordering results and performance analysis'. *IEEE Transactions on Wireless Communications*. 2013, vol. 12(10), pp. 5208–22.

[42]   Andrews J.G. 'Seven ways that HetNets are a cellular paradigm shift'. *IEEE Communications Magazine*. 2013, vol. 51(3), pp. 136–44.

[43]   Bjornson E., Sanguinetti L., Kountouris M. 'Deploying dense networks for maximal energy efficiency: small cells meet massive MIMO'. *IEEE Journal on Selected Areas in Communications*. 2016, vol. 34(4), pp. 832–47.

[44]   Larsson E.G., Edfors O., Tufvesson F., Marzetta T.L. 'Massive MIMO for next generation wireless systems'. *IEEE Communications Magazine*. 2014, vol. 52(2), pp. 186–95.

[45]   Ngo H.Q., Larsson E.G., Marzetta T.L. 'Energy and spectral efficiency of very large multiuser MIMO systems'. *IEEE Transactions on Communications*. 2013, vol. 61(4), pp. 1436–49.

[46]   Beibei Wang., Liu K.J.R. 'Advances in cognitive radio networks: a survey'. *IEEE Journal of Selected Topics in Signal Processing*. 2011, vol. 5(1), pp. 5–23.

[47]   Quan Z., Cui S., Sayed A.H. 'Optimal linear cooperation for spectrum sensing in cognitive radio networks'. *IEEE Journal of Selected Topics in Signal Processing*. 2008, vol. 2(1), pp. 28–40.

[48]   Nguyen V., Tran L., Duong T.Q., Shin O., Farrell R. 'An efficient precoder design for multiuser MIMO cognitive radio networks with interference constraints'. *IEEE Transactions on Vehicular Technology*. 2017, vol. 66(5), pp. 3991–4004.

[49]   Ngo H., Ashikhmin A., Yang H., Larsson E., Marzetta T. 'Cellfree massive MIMO versus small cells'. *IEEE Transactions on Wireless Communication*. 2015, vol. 99, pp. 1834–50.

[50]   Nayebi E., Ashikhmin A., Marzetta T.L., Yang H. 'Cell-free massive MIMO systems'. Proceedings of the 49th Asilomar Conference on Signals, Systems and Computers, California, USA; Nov 2015. pp. 695–9.

[51]   Nguyen L.D., Duong T.Q., Ngo H.Q., Tourki K. 'Energy efficiency in cell-free massive MIMO with zero-forcing precoding design'. *IEEE Communications Letters*. 2017, vol. 21(8), pp. 1871–4.

[52]   Rappaport T.S., Sun S., Mayzus R., *et al.* 'Millimeter wave mobile communications for 5G cellular: it will work!' *IEEE Access*. 2013, vol. 1, pp. 335–49.

[53]   Ghosh A., Thomas T.A., Cudak M.C., *et al.* 'Millimeter-wave enhanced local area systems: a high-data-rate approach for future wireless networks'. *IEEE Journal on Selected Areas in Communications*. 2014, vol. 32(6), pp. 1152–63.

[54]   Roh W., Seol J.-Y., Park J., *et al.* 'Millimeter-wave beamforming as an enabling technology for 5G cellular communications: theoretical feasibility and prototype results'. *IEEE Communications Magazine*. 2014, vol. 52(2), pp. 106–13.

[55]   Priya S., Inman D.J. *Energy harvesting technologies*. 21. Springer; 2009.

[56]   Lu X., Wang P., Niyato D., Kim D.I., Han Z. 'Wireless networks with RF energy harvesting: a contemporary survey'. *IEEE Communications Surveys & Tutorials*. 2015, vol. 17(2), pp. 757–89.

[57] Tam H.H.M., Tuan H.D., Nasir A.A., Duong T.Q., Poor H.V. 'MIMO energy harvesting in full-duplex multi-user networks'. *IEEE Transactions on Wireless Communications*. 2017, vol. 16(5), pp. 3282–97.

[58] Nasir A.A., Tuan H.D., Duong T.Q., Poor H.V. 'Secrecy rate beamforming for multicell networks with information and energy harvesting'. *IEEE Transactions on Signal Processing*. 2017, vol. 65(3), pp. 677–89.

[59] Nasir A.A., Tuan H.D., Duong T.Q., Poor H.V. 'Secure and energy-efficient beamforming for simultaneous information and energy transfer'. *IEEE Transactions on Wireless Communications*. 2017, vol. 16(11), pp. 7523–37.

[60] Yang M., Guo D., Huang Y., Duong T.Q., Zhang B. 'Physical layer security with threshold-based multiuser scheduling in multi-antenna wireless networks'. *IEEE Transactions on Communications*. 2016, vol. 64(12), pp. 5189–202.

[61] Fan L., Yang N., Duong T.Q., Elkashlan M., Karagiannidis G.K. 'Exploiting direct links for physical layer security in multiuser multirelay networks'. *IEEE Transactions on Wireless Communications*. 2016, vol. 15(6), pp. 3856–67.

[62] Jameel F., Wyne S., Kaddoum G., Duong T.Q. 'A comprehensive survey on cooperative relaying and jamming strategies for physical layer security'. *IEEE Communications Surveys Tutorials*. 2018, p. 1.

[63] Dinkelbach W. 'On nonlinear fractional programming'. *Management Science*. 1967, vol. 13(7), pp. 492–8.

[64] Yan Y., Qian Y., Sharif H., Tipper D. 'A survey on smart grid communication infrastructures: motivations, requirements and challenges'. *IEEE Communications Surveys & Tutorials*, vol. 15(1), pp. 5–20.

[65] Buzzi S., I C.-L., Klein T.E., Poor H.V., Yang C., Zappone A. 'A survey of energy-efficient techniques for 5G networks and challenges ahead'. *IEEE Journal on Selected Areas in Communications*. 2016, vol. 34(4), pp. 697–709.

[66] Andrews J.G., Buzzi S., Choi W., *et al.* 'What will 5G be?' *IEEE Journal on Selected Areas in Communications*. 2014, vol. 32(6), pp. 1065–82.

[67] GreenTouch Green Meter Research Study. *Reducing the net energy consumption in communications networks by up to 90% by 2020 (version 1.0)* [online]. 2013. Available from https://s3-us-west-2.amazonaws.com/bell-labsmicrosite- greentouch/ [Accessed 15/06/2021].

[68] Fehske A., Fettweis G., Malmodin J., Biczok G. 'The global footprint of mobile communications: the ecological and economic perspective'. *IEEE Communications Magazine*. 2011, vol. 49(8), pp. 55–62.

[69] Zappone A., Sanguinetti L., Bacci G., Jorswieck E., Debbah M. 'Energy-efficient power control: a look at 5G wireless technologies'. *IEEE Transactions on Signal Processing*. 2016, vol. 64(7), pp. 1668–83.

[70] Mattai J., Joseph M. *Real-time systems: specification, verification, and analysis*. Prentice Hall PTR; 1995.

[71] Kopetz H. *Real-time systems: design principles for distributed embedded applications*. Springer Science & Business Media; 2011.

[72] Laplante P.A. *Real-time systems design and analysis*. 3. New York: Wiley; 2004.

[73] ERCIM.. ' 'Special: embedded systems". *European Research Consortium for Informatics and Mathematics (ERCIM)*. 2003.

[74] Mattingley J., Boyd S. 'Real-time convex optimization in signal processing'. *IEEE Signal Processing Magazine*. 2010, vol. 27(3), pp. 50–61.

[75] Spall J.C. *Stochastic optimization Handbook of Computational Statistics*. Springer; 2012. pp. 173–201.

[76] Ali A., Kolter J.Z., Diamond S., Boyd S. 'Disciplined convex stochastic programming: a new framework for stochastic optimization'. Proceedings of the Thirty-First Conference on Uncertainty in Artificial Intelligence; Arlington, Virginia, United States; 2015. pp. 62–71.

[77] Ribeiro A. 'Ergodic stochastic optimization algorithms for wireless communication and networking'. *IEEE Transactions on Signal Processing*. 2010, vol. 58(12), pp. 6369–86.

[78] Collet P., Rennard J.-P. 'Stochastic optimization algorithms'. *arXiv preprint arXiv:0704.3780*. 2007.

[79] Bergstra J., Bengio Y. 'Random search for hyper-parameter optimization'. *Journal of Machine Learning Research*. 2012, vol. 13(Feb), pp. 281–305.

[80] Aarts E.H.L., Korst J.H.M. *Simulated annealing and Boltzmann machines a stochastic approach to combinatorial optimization and neural computing*. Chichester: Wiley; 1989.

[81] Back T. *Evolutionary algorithms in theory and practice: evolution strategies, evolutionary programming, genetic algorithms*. Oxford University Press; 1996.

[82] Gen M., Cheng R. *Genetic algorithms and engineering optimization*. 7. John Wiley & Sons; 2000.

[83] Kennedy J. *Particle swarm optimization. Encyclopedia of machine learning*. Springer; 2011. pp. 760–6.

[84] Shi Y. 'Particle swarm optimization: developments, applications and resources'. 2001 Proceedings of the 2001 Congress on in Evolutionary Computation, vol. 1. IEEE; 2001. pp. 81–6.

[85] Dorigo M., Di Caro G. 'Ant colony optimization: a new metaheuristic'. Proceedings of the 1999 Congress on in Evolutionary Computation, 1999. CEC 99. vol. 2. IEEE; 1999. pp. 1470–7.

[86] Dorigo M., Birattari M. *Ant colony optimization Encyclopedia of Machine Learning*. Springer; 2011. pp. 36–9.

[87] Karaboga D., Basturk B. 'A powerful and efficient algorithm for numerical function optimization: artificial bee colony (ABC) algorithm'. *Journal of Global Optimization*. 2007, vol. 39(3), pp. 459–71.

[88] Gilks W.R., Richardson S., Spiegelhalter D. *Markov chain Monte Carlo in practice*. CRC Press; 1995.

[89] Zhang F., Cui Y., Lau V.K.N., Liu A. 'Distributed stochastic optimization for weakly coupled systems with applications to wireless communications'. *CoRR*. 2016, vol. abs/1606.07606.

[90] Zheng G., Zhang Y., Ji C., Wong K.-K. 'A stochastic optimization approach for joint relay assignment and power allocation in orthogonal amplify-and-forward cooperative wireless networks'. *IEEE Transactions on Wireless Communications*. 2011, vol. 10(12), pp. 4091–9.

[91] Lee J.-H., Lee J.-Y. 'Optimal beamforming-selection spatial precoding using population-based stochastic optimization for massive wireless MIMO communication systems'. *Journal of the Franklin Institute*. 2017, vol. 354(10), pp. 4247–72.

[92] Neely M. *Stochastic network optimization with application to communication and queueing systems. Morgan & claypool [online]*. 2010. Available from http://ieeexplore.ieee.org/xpl/articleDetails.jsp? arnumber=6813406 [Accessed 15/06/2021].

[93] Gould N., Orban D., Toint P. 'Numerical methods for large-scale nonlinear optimization'. *Acta Numerica*. 2005, vol. 14, pp. 299–361.

[94] Cevher V., Becker S., Schmidt M. 'Convex optimization for big data: scalable, randomized, and parallel algorithms for big data analytics'. *IEEE Signal Processing Magazine*. 2014, vol. 31(5), pp. 32–43.

[95] Gondzio J., Gruca J.A., Hall J.A.J., Laskowski W., Żukowski M. 'Solving large-scale optimization problems related to Bell's Theorem'. *Journal of Computational and Applied Mathematics*. 2014, vol. 263, pp. 392–404.

[96] D.-Z. D., Pardalos P.M. *Handbook of combinatorial optimization: supplement*. 1. Springer Science & Business Media; 2013.

[97] Nguyen L., Kortun A., Duong T. 'An introduction of real-time embedded optimisation programming for UAV systems under disaster communication'. *EAI Endorsed Transactions on Industrial Networks and Intelligent Systems*. 2018, vol. 5(17) 156080.

[98] Bertsekas D.P. *Nonlinear programming*. Belmont: Athena scientific Belmont; 1999.

[99] Juditsky A., Nemirovski A. 'First order methods for nonsmooth convex large-scale optimization, I: general purpose methods'. *Optimization for Machine Learning, MIT Press*. 2011, pp. 1–28.

[100] Fletcher R. *Practical methods of optimization*. John Wiley & Sons; 2013.

[101] Kingma D.P., Ba J. 'Adam: a method for stochastic optimization'. *arXiv preprint arXiv:1412.6980*. 2014.

[102] Boyd S., Parikh N., Chu E., Peleato B., Eckstein J. 'Distributed optimization and statistical learning via the alternating direction method of multipliers'. *Foundations and Trends R in Machine Learning*. 2011, vol. 3(1), pp. 1–122.

[103] Rumelhart D.E., McClelland J.L., Group P.R. *Parallel distributed processing*. 1. Cambridge, MA: MIT Press; 1987.

[104] Grama A. *Introduction to parallel computing*. Pearson Education; 2003.

[105] Migdalas A., Pardalos P.M., Storøy S. *Parallel computing in optimization*. 7. Springer Science & Business Media; 2013.

[106] Bhutani G. 'Application of machine-learning based prediction techniques in wireless networks'. *International Journal of Communications, Network and System Sciences*. 2014, vol. 07(05) 131–40.

[107]   Alsheikh M.A., Lin S., Niyato D., Tan H.-P. 'Machine learning in wireless sensor networks: algorithms, strategies, and applications'. *IEEE Communications Surveys & Tutorials.* 2014, vol. 16(4), pp. 1996–2018.

[108]   Jiang C., Zhang H., Ren Y., Han Z., Chen K.-C., Hanzo L. 'Machine learning paradigms for next-generation wireless networks'. *IEEE Wireless Communications.* 2017, vol. 24(2), pp. 98–105.

[109]   Klaine P.V., Imran M.A., Onireti O., Souza R.D. 'A survey of machine learning techniques applied to self-organizing cellular networks'. *IEEE Communications Surveys & Tutorials.* 2017, vol. 19(4), pp. 2392–431.

[110]   Chen M., Challita U., Saad W., Yin C., Debbah M. 'Machine learning for wireless networks with artificial intelligence: a tutorial on neural networks'. *arXiv preprint arXiv:1710.02913,.* 2017.

[111]   van Lenthe E., Ehlers A., Baerends E.-J. 'Geometry optimizations in the zero order regular approximation for relativistic effects'. *The Journal of Chemical Physics.* 1999, vol. 110(18), pp. 8943–53.

[112]   Duchi J.C., Jordan M.I., Wainwright M.J., Wibisono A. 'Optimal rates for zero-order convex optimization: the power of two function evaluations'. *IEEE Transactions on Information Theory.* 2015, vol. 61(5), pp. 2788–806.

[113]   Nikias C.L. 'Higher-order spectral analysis'. 1993 Proceedings of the 15th Annual International Conference of the IEEE in Engineering in Medicine and Biology Society, IEEE; San Diego, CA, USA, 31 Oct; 1993. p. 319.

[114]   Chambolle A., Pock T. 'A first-order primal-dual algorithm for convex problems with applications to imaging'. *Journal of mathematical imaging and vision.* 2011, vol. 40(1), pp. 120–45.

[115]   Powell M.J. *A fast algorithm for nonlinearly constrained optimization calculations. Numerical Analysis.* Springer; 1978. pp. 144–57.

[116]   Beck A. *First-order methods in optimization.* 25. SIAM; 2017. pp. 1–467.

[117]   Amann H. 'Fixed point equations and nonlinear eigenvalue problems in ordered banach spaces'. *SIAM Review.* 1976, vol. 18(4), pp. 620–709.

[118]   Gurevich Y., Shelah S. 'Fixed-point extensions of first-order logic'. *Annals of Pure and Applied Logic.* 1986, vol. 32, pp. 265–80.

[119]   d'Aspremont A., Banerjee O., El Ghaoui L. 'First-order methods for sparse covariance selection'. *SIAM Journal on Matrix Analysis and Applications.* 2008, vol. 30(1), pp. 56–66.

[120]   Wang Y., Yang J., Yin W., Zhang Y. 'A new alternating minimization algorithm for total variation image reconstruction'. *SIAM Journal on Imaging Sciences.* 2008, vol. 1(3), pp. 248–72.

[121]   Zinkevich M., Weimer M., Li L., Smola A.J. 'Parallelized stochastic gradient descent'. Advances in neural information processing systems; 2010. pp. 2595–603.

[122]   Richtarik P., Schmidt M. 'Modern convex optimization methods for large-scale empirical risk minimization'. *Tutorial in ICML.* 2015, pp. 1–282.

[123]   Jordan M. 'On gradient-based optimization: accelerated, distributed, asynchronous and stochastic'. Proceedings of the 2017 ACM SIGMETRICS/

International Conference on Measurement and Modeling of Computer Systems. ACM; 2017. pp. 58–58.

[124] Beck A., Teboulle M. 'A fast iterative shrinkage-thresholding algorithm for linear inverse problems'. *SIAM Journal on Imaging Sciences*. 2009, vol. 2(1), pp. 183–202.

[125] Parikh N. 'Proximal algorithms'. *Foundations and Trends® in Optimization*. 2014, vol. 1(3), pp. 127–239.

[126] Nesterov Y. 'Smooth minimization of non-smooth functions'. *Mathematical Programming*. 2005, vol. 103(1), pp. 127–52.

[127] Combettes P.L., Pesquet J.-C. 'Proximal splitting methods in signal processing'. *Fixed-Point Algorithms for Inverse Problems in Science and Engineering*. Springer; 2011. pp. 185–212.

[128] Farivar M. 'Accelerated proximal-gradient method for large scale convex problems'. 2015.

[129] Jiang K., Sun D., Toh K.-C. 'An inexact accelerated proximal gradient method for large scale linearly constrained convex SDP'. *SIAM Journal on Optimization*. 2012, vol. 22(3), pp. 1042–64.

[130] Bottou L. 'Large-scale machine learning with stochastic gradient descent'. *Proceedings of COMPSTAT'2010*. Springer; 2010. pp. 177–86.

[131] Zhang T. Solving large scale linear prediction problems using stochastic gradient descent algorithms. Proceedings of the Twenty-First International Conference on Machine Learning. ACM; 2004. p. 116.

[132] Marbach P., Berry R. 'Downlink resource allocation and pricing for wireless networks'. Twenty-First Annual Joint Conference of the IEEE Computer and Communications Societies. Proceedings. IEEE INFOCOM 2002. vol. 3. IEEE; 2002. pp. 1470–9.

[133] Pahalawatta P., Berry R., Pappas T., Katsaggelos A. 'Content-aware resource allocation and packet scheduling for video transmission over wireless networks'. *IEEE Journal on Selected Areas in Communications*. 2007, vol. 25(4) 749–59.

[134] Lin X., Shroff N.B. The impact of imperfect scheduling on crosslayer rate control in wireless networks. 24th Annual Joint Conference of the IEEE Computer and Communications Societies. Proceedings IEEE INFOCOM 2005. vol. 3. IEEE; 2005. pp. 1804–14.

[135] Xiaojun Lin., Shroff N.B. 'The impact of imperfect scheduling on cross-layer congestion control in wireless networks'. *IEEE/ACM Transactions on Networking*. 2006, vol. 14(2), pp. 302–15.

[136] Hua C., T.-S.P.Yum. 'Optimal routing and data aggregation for maximizing lifetime of wireless sensor networks'. *IEEE/ACM Transactions on Networking*. 2008, vol. 16(4), pp. 892–903.

[137] Kim S.-J., Giannakis G.B. 'Optimal resource allocation for MIMO AD hoc cognitive radio networks'. *IEEE Transactions on Information Theory*. 2011, vol. 57(5), pp. 3117–31.

[138]   Cadambe V.R., Jafar S.A., Wang C. 'Interference alignment with asymmetric complex signaling—settling the Høst-Madsen-Nosratinia conjecture'. *IEEE Transactions on Information Theory*. 2010, vol. 56(9), pp. 4552–65.

[139]   Shi Q., Razaviyayn M., Luo Z.-Q., He C. 'An iteratively weighted MMSE approach to distributed sum-utility maximization for a MIMO interfering broadcast channel'. *IEEE Transactions on Signal Processing*. 2011, vol. 59(9), pp. 4331–40.

[140]   Razaviyayn M., Sanjabi M., Luo Z.-Q. 'Linear transceiver design for interference alignment: complexity and computation'. *IEEE Transactions on Information Theory*. 2012, vol. 58(5), pp. 2896–910.

[141]   Kim D., Torlak M. 'Optimization of interference alignment beamforming vectors'. *IEEE Journal on Selected Areas in Communications*. 2010, vol. 28(9), pp. 1425–34.

[142]   Nasir A.A., Tuan H.D., Duong T.Q., Poor H.V. 'Secure and energy-efficient beamforming for simultaneous information and energy transfer'. *IEEE Transactions on Wireless Communications*. 2017, vol. 16(11), pp. 7523–37.

[143]   Nghia N.T., Tuan H.D., Duong T.Q., Poor H.V. 'MIMO beamforming for secure and energy-efficient wireless communication'. *IEEE Signal Processing Letters*. 2017, vol. 24(2), pp. 236–9.

[144]   Tam H.H.M., Tuan H.D., Nasir A.A., Duong T.Q., Poor H.V. 'MIMO energy harvesting in full-duplex multi-user networks'. *IEEE Transactions on Wireless Communications*. 2017, vol. 16(5), pp. 3282–97.

[145]   Nasir A.A., Tuan H.D., Ngo D.T., Duong T.Q., Poor H.V. 'Beamforming design for wireless information and power transfer systems: receive power-splitting versus transmit time-switching'. *IEEE Transactions on Communications*. 2017, vol. 65(2), pp. 876–89.

[146]   Nasir A.A., Tuan H.D., Duong T.Q., Poor H.V. 'Secrecy rate beamforming for multicell networks with information and energy harvesting'. *IEEE Transactions on Signal Processing*. 2017, vol. 65(3), pp. 677–89.

[147]   Nguyen L.D., Tuan H.D., Duong T.Q., Poor H.V. 'Beamforming and power allocation for energy-efficient massive MIMO'. *22th International Conference on Digital Signal Processing and Its Applications*. 2017, vol. 105, pp. 2165–3577.

[148]   Nguyen L.D., Tuan H.D., Duong T.Q., Dobre O.A., Poor H.V. 'Downlink beamforming for energy-efficient heterogeneous networks with massive MIMO and small cells'. *IEEE Transactions on Wireless Communications*. 2017.

[149]   Nguyen L.D., Tuan H.D., Duong T.Q., Poor H.V. 'Multicell massive MIMO beamforming in assuring QoS for large numbers of users'. *arXiv preprint arXiv:1712.03548*. 2017.

[150]   Nguyen L.D., Duong T.Q., Nguyen D.N., Tran L.N. Energy efficiency maximization for heterogeneous networks: a joint linear precoder design and small-cell switching-off approach. *IEEE Global Conference on Signal and Information Processing*; Washington, DC, USA; 2016.

[151] Kopetz H. *Real-time systems: design principles for distributed embedded applications*. Springer Science & Business Media; 2011.

[152] Bertsekas D.P., Tsitsiklis J.N. *Parallel and distributed computation: numerical methods*. 23. Englewood Cliffs, NJ: Prentice Hall; 1989.

[153] Schnabel R.B. *Parallel computing in optimization. Computational mathematical programming*. Springer; 1985. pp. 357–81.

[154] Deb K. *Multi-objective optimization. Search Methodologies*. Springer; 2014. pp. 403–49.

[155] Marler R.T., Arora J.S. 'Survey of multi-objective optimization methods for engineering'. *Structural and Multidisciplinary Optimization*. 2004, vol. 26(6), pp. 369–95.

[156] Coello C.A.C., Lamont G.B., Van Veldhuizen D.A. *Evolutionary algorithms for solving multi-objective problems*. 5. Springer; 2007.

[157] Deb K. *Multi-objective optimization using evolutionary algorithms*. 16. John Wiley & Sons; 2001.

[158] Konak A., Coit D.W., Smith A.E. 'Multi-objective optimization using genetic algorithms: a tutorial'. *Reliability Engineering & System Safety*. 2006, vol. 91(9), pp. 992–1007.

[159] Deb K., Thiele L., Laumanns M., Zitzler E. 'Scalable multiobjective optimization test problems'. CEC'02 Proceedings of the 2002 Congress on Evolutionary Computation, vol. 1. IEEE; 2002. pp. 825–30.

[160] Scutari G., Facchinei F., Lampariello L., Song P. 'Parallel and distributed methods for nonconvex optimization'. 2014 IEEE International Conference on Acoustics, Speech and Signal Processing (ICASSP). IEEE; 2014. pp. 840–4.

[161] Boyd S., Parikh N., Chu E., Peleato B., Eckstein J. 'Distributed optimization and statistical learning via the alternating direction method of multipliers'. *Foundations and Trends® in Machine Learning*. 2010, vol. 3(1), pp. 1–122.

[162] O'Donoghue B., Stathopoulos G., Boyd S. 'A splitting method for optimal control'. *IEEE Transactions on Control Systems Technology*. 2013, vol. 21(6), pp. 2432–42.

[163] Peng Z., Yan M., Yin W. 'Parallel and distributed sparse optimization'. 2013 Asilomar Conference on Signals, Systems and Computers. IEEE; 2013–659–646.

[164] Deng W., Lai M.-J., Peng Z., Yin W. 'Parallel multi-block ADMM with o(1 / k) convergence'. *Journal of Scientific Computing*. 2017, vol. 71(2), pp. 712–36.

[165] Hao N., Oghbaee A., Rostami M., Derbinsky N., Bento J. 'Testing fine-grained parallelism for the admm on a factor-graph'. *Parallel and Distributed Processing Symposium Workshops, 2016 IEEE International*. IEEE; 2016. pp. 835–44.

[166] Schutte J.F., Reinbolt J.A., Fregly B.J., Haftka R.T., George A.D. 'Parallel global optimization with the particle swarm algorithm'. *International Journal for Numerical Methods in Engineering*. 2004, vol. 61(13), pp. 2296–315.

[167]   Abramov O., Katueva Y. 'Application of parallel computing techniques for stochastic optimization problems'. *2004 5th Asian Control Conference. vol. 1. IEEE*; 2004. pp. 435–41.

[168]   Recht B., Ré C. 'Parallel stochastic gradient algorithms for large-scale matrix completion'. *Mathematical Programming Computation*. 2013, vol. 5(2), pp. 201–26.

[169]   Facchinei F., Scutari G., Sagratella S. 'Parallel selective algorithms for nonconvex big data optimization'. *IEEE Transactions on Signal Processing*. 2015, vol. 63(7), pp. 1874–89.

[170]   Villegas F.J., Cwik T., Rahmat-Samii Y., Manteghi M. 'A parallel electromagnetic genetic-algorithm optimization (EGO) application for patch antenna design'. *IEEE Transactions on Antennas and Propagation*. 2004, vol. 52(9), pp. 2424–35.

[171]   Jin N., Rahmat-Samii Y. 'Parallel particle swarm optimization and finite-difference time-domain (PSO/FDTD) algorithm for multiband and wide-band patch antenna designs'. *IEEE Transactions on Antennas and Propagation*. 2005, vol. 53(11), pp. 3459–68.

[172]   Kosta S., Aucinas A., Hui P., Mortier R., Zhang X. 'Thinkair: dynamic resource allocation and parallel execution in the cloud for mobile code offloading'. *2012 Proceedings IEEE INFOCOM. IEEE*. 2012, pp. 945–53.

[173]   Palomar D.P., Mung Chiang, Chiang M. 'A tutorial on decomposition methods for network utility maximization'. *IEEE Journal on Selected Areas in Communications*. 2006, vol. 24(8), pp. 1439–51.

[174]   Yonghoon Choi., Hoon Kim., Sang-wook Han., Youngnam Han. 'Joint resource allocation for parallel multi-radio access in heterogeneous wireless networks'. *IEEE Transactions on Wireless Communications*. 2010, vol. 9(11), pp. 3324–9.

[175]   Predd J.B., Kulkarni S.B., Poor H.V. 'Distributed learning in wireless sensor networks'. *IEEE Signal Processing Magazine*. 2006, vol. 23(4), pp. 56–69.

[176]   Basagni S., Conti M., Giordano S., Stojmenovic I. *Mobile ad hoc networking*. John Wiley & Sons; 2004.

[177]   Zeng X., Bagrodia R., Gerla M. 'Glomosim: a library for parallel simulation of large-scale wireless networks'. *ACM SIGSIM Simulation Digest, IEEE Computer Society*. 1998, vol. 28(1), pp. 154–61.

[178]   Goodfellow I., Bengio Y., Courville A., Bengio Y. *Deep learning*. 1. Cambridge: MIT Press; 2016.

[179]   LeCun Y., Bengio Y., Hinton G. 'Deep learning'. *Nature*. 2015, vol. 521(7553), pp. 436–44.

[180]   Deng L. 'Deep learning: methods and applications'. *Foundations and Trends® in Signal Processing*. 2013, vol. 7(3–4), pp. 197–387.

[181]   Schmidhuber J. 'Deep learning in neural networks: an overview'. *Neural Networks*. 2015, vol. 61, pp. 85–117.

[182]   Kibria M.G., Nguyen K., Villardi G.P., Ishizu K., Kojima F. 'Big data analytics and artificial intelligence in next-generation wireless networks'. *arXiv preprint arXiv:1711.10089*. 2017.

[183]   Li R., Zhao Z., Zhou X., *et al.* 'Intelligent 5G: when cellular networks meet artificial intelligence'. *IEEE Wireless Communications*. 2017, vol. 24(5), pp. 175–83.

[184]   Barr A., Feigenbaum E.A. *The handbook of artificial intelligence*. 2. Butterworth-Heinemann; 2014.

[185]   Bishop C.M. *Pattern recognition and machine learning*. Springer; 2006.

[186]   Alpaydin E. *Introduction to machine learning*. MIT Press; 2009.

[187]   Shalev-Shwartz S., Ben-David S. *Understanding machine learning: from theory to algorithms*. Cambridge University Press; 2014.

[188]   LeCun Y., Bengio Y., Hinton G. 'Deep learning'. *Nature*. 2015, vol. 521(7553), pp. 436–44.

[189]   Aliu O.G., Imran A., Imran M.A., Evans B. 'A survey of self organisation in future cellular networks'. *IEEE Communications Surveys & Tutorials*. 2013, vol. 15(1), pp. 336–61.

[190]   Poor R.D. *Self-organizing network*. US Patent 6,028,857; 2000.

[191]   Vesanto J., Alhoniemi E. 'Clustering of the self-organizing map'. *IEEE Transactions on Neural Networks*. 2000, vol. 11(3), pp. 586–600.

[192]   Ramiro J., Hamied K. *Self-organizing networks (SON): self-planning, self-optimization and self-healing for GSM, UMTS and LTE*. John Wiley & Sons; 2011.

[193]   Latif S., Pervez F., Usama M., Qadir J. 'Artificial intelligence as an enabler for cognitive self-organizing future networks'. *arXiv preprint arXiv:1702.02823*. 2017.

[194]   Bkassiny M., Li Y., Jayaweera S.K. 'A survey on machine-learning techniques in cognitive radios'. *IEEE Communications Surveys & Tutorials*. 2013, vol. 15(3), pp. 1136–59.

[195]   Zorzi M., Zanella A., Testolin A., De Filippo De Grazia M., Zorzi M. 'Cognition-based networks: a new perspective on network optimization using learning and distributed intelligence'. *IEEE Access*. 2015, vol. 3, pp. 1512–30.

[196]   Haykin S. 'Cognitive radio: brain-empowered wireless communications'. *IEEE Journal on Selected Areas in Communications*. 2005, vol. 23(2), pp. 201–20.

[197]   Ovalle D., Restrepo D., Montoya A. *Artificial intelligence for wireless sensor networks enhancement. Smart Wireless Sensor Networks*. InTech; 2010.

[198]   Kulkarni R.V., Forster A., Venayagamoorthy G.K. 'Computational intelligence in wireless sensor networks: a survey'. *IEEE Communications Surveys & Tutorials*. 2011, vol. 13(1), pp. 68–96.

[199]   Li R., Zhao Z., Zhou X., *et al.* 'Intelligent 5G: when cellular networks meet artificial intelligence'. *IEEE Wireless Communications*. 2017, vol. 24(5), pp. 175–83.

[200]   Bennett K.P., Parrado-Hernández E. 'The interplay of optimization and machine learning research'. *Journal of Machine Learning Research*. 2006, vol. 7(Jul), pp. 1265–81.

[201]   Sra S., Nowozin S., Wright S.J. *Optimization for machine learning*. MIT Press; 2012.

[202]   Snoek J., Larochelle H., Adams R.P. 'Practical Bayesian optimization of machine learning algorithms'. Advances in Neural Information Processing Systems; 2012. pp. 2951–9.

[203]   Scholkopf B., Smola A.J. *Learning with kernels: support vector machines, regularization, optimization, and beyond*. MIT Press; 2001.

[204]   Kennedy J. 'Particle swarm optimization'. *Encyclopedia of Machine Learning*. Springer; 2011. pp. 760–6.

[205]   Le Q.V., Ngiam J., Coates A., Lahiri A., Prochnow B., Ng A.Y. 'On optimization methods for deep learning'. Proceedings of the 28th International Conference on International Conference on Machine Learning. Omnipress; 2011. pp. 265–72.

[206]   Gao J., Jamidar R. 'Machine learning applications for data center optimization'. Google White Paper; 2014.

[207]   Y. C. M. W.H.S., Colmenarejo G. 'Learning to learn for global optimization of black box functions'. *Stat*. 2016, vol. 1050,11.

[208]   Li K., Malik J. 'Learning to optimize'. *arXiv preprint arXiv:1606.01885*. 2016.

[209]   Sun H., Chen X., Shi Q., Hong M., Fu X., Sidiropoulos N.D. 'Learning to optimize: training deep neural networks for wireless resource management'. *arXiv preprint arXiv:1705.09412*. 2017.

[210]   Li K., Malik J. 'Learning to optimize neural nets'. *arXiv preprint arXiv:1703.00441*. 2017.

[211]   Cochocki A., Unbehauen R. *Neural networks for optimization and signal processing*. John Wiley & Sons, Inc; 1993.

[212]   Cassioli A., Di Lorenzo D., Locatelli M., Schoen F., Sciandrone M. 'Machine learning for global optimization'. *Computational Optimization and Applications*. 2012, vol. 51(1), pp. 279–303.

[213]   Lee Y.-J., Mangasarian O.L. 'SSVM: a smooth support vector machine for classification'. *Computational Optimization and Applications*. 2001, vol. 20(1), pp. 5–22.

[214]   Chapelle O. 'Training a support vector machine in the primal'. *Neural Computation*. 2007, vol. 19(5), pp. 1155–78.

[215]   Sutton R.S., Barto A.G. *Reinforcement learning: an introduction*. 1. Cambridge: MIT press; 1998.

[216]   Kaelbling L.P., Littman M.L., Moore A.W. 'Reinforcement learning: a survey'. *Journal of Artificial Intelligence Research*. 1996, vol. 4, pp. 237–85.

[217]   Hecht-Nielsen R. 'Theory of the backpropagation neural network'. *Neural Networks for Perception*. Elsevier; 1992. pp. 65–93.

[218]   M.-H., Nguyen T., Garcia-Palacios E., *et al.* 'Spectrum-sharing UAV-assisted missioncritical communication: learning-aided real-time optimisation'. *IEEE Access: Practical Innovations, Open Solutions*. 2021, vol. 9, pp. 11622–32.

[219] Ruvolo P.L., Fasel I., Movellan J.R. 'Optimization on a budget: a reinforcement learning approach'. *Advances in Neural Information Processing Systems*; 2009. pp. 1385–92.

[220] Bernhard K., Vygen J. *Combinatorial optimization: theory and algorithms*. Third Edition. 2005. Springer; 2008.

[221] Grant M., Boyd S. *CVX: MATLAB software for disciplined convex programming, version 2.1 [online]*. 2014. Available from http://cvxr.com/cvx [Accessed 15/06/2021].

[222] Lofberg J. 'YALMIP: a toolbox for modeling and optimization in MATLAB'. *2004 IEEE International Conference on Robotics and Automation (IEEE Cat. No.04CH37508)*; 2004. pp. 284–9.

[223] Diamond S., Boyd S. 'CVXPY: a Python-embedded modeling language for convex optimization'. *Journal of Machine Learning Research : JMLR*. 2016, vol. 17(83), pp. 1–5.

[224] Fu A., Narasimhan B., Boyd S. 'CVXR: an R package for disciplined convex optimization'. *arXiv preprint arXiv:1711.07582*. 2017.

[225] Udell M., Mohan K., Zeng D., Hong J., Diamond S., Boyd S. 'Convex optimization in Julia'. *SC14 Workshop on High Performance Technical Computing in Dynamic Languages*; 2014.

[226] Dunning I., Huchette J., Lubin M. 'Jump: a modeling language for mathematical optimization'. *SIAM Review*. 2017, vol. 59(2), pp. 295–320.

[227] Frew E.W., Brown T.X. 'Airborne communication networks for small unmanned aircraft systems'. *Proceedings of the IEEE*. 2008, vol. 96(12) 2008–27.

[228] Gupta L., Jain R., Vaszkun G. 'Survey of important issues in UAV communication networks'. *IEEE Communications Surveys & Tutorials*. 2016, vol. 18(2), pp. 1123–52.

[229] Olsson P.-M., Kvarnström J., Doherty P., Burdakov O., Holmberg K. 'Generating UAV communication networks for monitoring and surveillance'. *2010 11th International Conference on Control Automation Robotics & Vision (ICARCV)*. IEEE; 2010. pp. 1070–7.

[230] Agogino A., HolmesParker C., Tumer K. 'Evolving large scale UAV communication system'. *Proceedings of the 14th Annual Conference on Genetic and Evolutionary Computation*. ACM; 2012. pp. 1023–30.

[231] Bekmezci İlker., Sahingoz O.K., Temel Şamil. 'Flying ad-hoc networks (FANETs): a survey'. *Ad Hoc Networks*. 2013, vol. 11(3), pp. 1254–70.

[232] Schoenwald D.A. 'Auvs: in space, air, water, and on the ground'. *IEEE Control Systems*. 2000, vol. 20(6), pp. 15–18.

[233] Romer K., Mattern F. 'The design space of wireless sensor networks'. *IEEE Wireless Communications*. 2004, vol. 11(6), pp. 54–61.

[234] Shima T., Rasmussen S. 'UAV cooperative decision and control: challenges and practical approaches'. *SIAM*. 2009.

[235] Zeng Y., Zhang R. 'Energy-efficient UAV communication with trajectory optimization'. *IEEE Transactions on Wireless Communications*. 2017, vol. 16(6), pp. 3747–60.

[236] Wu Q., Zeng Y., Zhang R. 'Joint trajectory and communication design for multi-UAV enabled wireless networks'. *IEEE Transactions on Wireless Communications*. 2018.

[237] Betts J.T. 'Survey of numerical methods for trajectory optimization'. *Journal of Guidance, Control, and Dynamics*. 1998, vol. 21(2), pp. 193–207.

[238] Ponda S., Kolacinski R., Frazzoli E. 'Trajectory optimization for target localization using small unmanned aerial vehicles'. AIAA Guidance, Navigation, and Control Conference; 2009. p. 6015.

[239] Bor-Yaliniz R.I., El-Keyi A., Yanikomeroglu H. 'Efficient 3-D placement of an aerial base station in next generation cellular networks'. *2016 IEEE International Conference on Communications*. 2016, pp. 1–5.

[240] Al-Hourani A., Kandeepan S., Jamalipour A. 'Modeling air-toground path loss for low altitude platforms in urban environments'. *2014 IEEE Global Communications Conference*. 2014, pp. 2898–904.

[241] Mozaffari M., Saad W., Bennis M., Debbah M. 'Efficient deployment of multiple unmanned aerial vehicles for optimal wireless coverage'. *IEEE Communications Letters*. 2016, vol. 20(8), pp. 1647–50.

[242] Al-Hourani A., Kandeepan S., Lardner S. 'Optimal LAP altitude for maximum coverage'. *IEEE Wireless Communications Letters*. 2014, vol. 3(6), pp. 569–72.

[243] Xu J., Zeng Y., Zhang R. 'UAV-enabled wireless power transfer: trajectory design and energy optimization'. *IEEE Transactions on Wireless Communications*. 2018, pp. 1–1.

[244] Vasan G., Singh A.K., Krishna K.M. Model predictive control for micro aerial vehicle systems (MAV) systems, CoRR, vol. abs/1412.2356. 2014. Available from http://arxiv.org/abs/1412. 2356.

[245] Bertrand S., Marzat J., Piet-Lahanier H., Kahn A., Rochefort Y. 'MPC strategies for cooperative guidance of autonomous vehicles'. *AerospaceLab*. 2014, vol. 8, pp. 1–18.

[246] Gavilan F., Vazquez R., Camacho E.F. 'An iterative model predictive control algorithm for UAV guidance'. *IEEE Transactions on Aerospace and Electronic Systems*. 2015, vol. 51(3), pp. 2406–19.

[247] Mase K. 'How to deliver your message from/to a disaster area'. *IEEE Communications Magazine*. 2011, vol. 49(1), pp. 52–7.

[248] Ali K., Nguyen H.X., Vien Q.-T., Shah P. 'Disaster management communication networks: challenges and architecture design'. 2015 IEEE International Conference on Pervasive Computing and Communication Workshops (PerCom Workshops). IEEE; 2015. pp. 537–42.

[249] Manoj B.S., Baker A.H. 'Communication challenges in emergency response'. *Communications of the ACM*. 2007, vol. 50(3), pp. 51–3.

[250] Lei S., Wang J., Chen C., Hou Y. 'Mobile emergency generator prepositioning and real-time allocation for resilient response to natural disasters'. *IEEE Transactions on Smart Grid*. 2018, vol. 9(3), pp. 2030–41.

[251]   Yuan W., Wang J., Qiu F., Chen C., Kang C., Zeng B. 'Robust optimization-based resilient distribution network planning against natural disasters'. *IEEE Transactions on Smart Grid*. 2016, vol. 7(6), pp. 2817–26.

[252]   Bianco A., Giraudo L., Hay D. 'Optimal resource allocation for disaster recovery'. 2010 IEEE Global Telecommunications Conference; 2010. pp. 1–5.

[253]   Lorincz K., Malan D.J., Fulford-Jones T.R.F., *et al.* 'Sensor networks for emergency response: challenges and opportunities'. *IEEE Pervasive Computing*. 2004, vol. 3(4), pp. 16–23.

[254]   N.-S. V., Masaracchia A., Nguyen L.D., Huynh B.-C. 'Natural disaster and environmental monitoring system for smart cities: design and installation insights'. *EAI Endorsed Transactions on Industrial Networks and Intelligent Systems*. 2018, vol. 5(16), p. 11.

[255]   Vo N., Duong T.Q., Guizani M. 'Quality of sustainability optimization design for mobile ad hoc networks in disaster areas'. 2015 IEEE Global Communications Conference; 2015. pp. 1–6.

[256]   Liu X., Li Z., Zhao N. 'Transceiver design and multi-hop D2D for UAV IoT coverage in disasters'. *IEEE Internet of Things Journal*. 2018, pp. 1–1.

[257]   Hayajneh A.M., Zaidi S.A.R., McLernon D.C., Di Renzo M., Ghogho M. 'Performance analysis of UAV enabled disaster recovery networks: a stochastic geometric framework based on cluster processes'. *IEEE Access*. 2018, vol. 6, pp. 26215–30.

[258]   Liu X., Ansari N. 'Resource allocation in UAV-assisted M2M communications for disaster rescue'. *IEEE Wireless Communications Letters*. 2018, pp. 1–1.

[259]   Nguyen M., Nguyen L.D., Duong T.Q., Tuan H.D. 'Realtime optimal resource allocation for embedded UAV communication systems'. *IEEE Wireless Communications Letters*. 2018, pp. 1–1.

[260]   Merwaday A., Tuncer A., Kumbhar A., Guvenc I. 'Improved throughput coverage in natural disasters: Unmanned aerial base stations for public-safety communications'. *IEEE Vehicular Technology Magazine*. 2016, vol. 11(4), pp. 53–60.

[261]   Zhang S., Liu J. 'Analysis and optimization of multiple unmanned aerial vehicle-assisted communications in post-disaster areas'. *IEEE Transactions on Vehicular Technology*. 2018, vol. 67(12), pp. 12049–60.

[262]   Erdelj M., Natalizio E., Chowdhury K.R., Akyildiz I.F. 'Help from the sky: Leveraging UAVs for disaster management'. *IEEE Pervasive Computing*. 2017, vol. 16(1), pp. 24–32.

[263]   Hossein Motlagh N., Taleb T., Arouk O. 'Low-altitude unmanned aerial vehicles-based internet of things services: comprehensive survey and future perspectives'. *IEEE Internet of Things Journal*. 2016, vol. 3(6), pp. 899–922.

[264]   Zhang S., Zhang H., He Q., Bian K., Song L. 'Joint trajectory and power optimization for UAV relay networks'. *IEEE Communications Letters: A Publication of the IEEE Communications Society*. 2017, vol. 99, p. 1.

[265]   Baek J., Han S.I., Han Y. 'Optimal resource allocation for nonorthogonal transmission in UAV relay systems'. *IEEE Wireless Communications Letters*. 2017, vol. 99, p. 1.

[266]   Zeng Y., Zhang R. 'Energy-efficient UAV communication with trajectory optimization'. *IEEE Transactions on Wireless Communications*. 2017, vol. 16(6), pp. 3747–60.

[267]   Bi S., Ho C.K., Zhang R. 'Wireless powered communication: opportunities and challenges'. *IEEE Communications Magazine*. 2015, vol. 53(4), pp. 117–25.

[268]   H.Wang., J.Wang., Ding G., L.Wang., Tsiftsis T.A., Sharma P.K. 'Resource allocation for energy harvesting-powered D2d communication underlaying UAV-assisted networks'. *IEEE Transactions on Green Communications and Networking*. 2017, vol. 99, p. 1.

[269]   Nasir A.A., Tuan H.D., Ngo D.T., Duong T.Q., Poor H.V. 'Beamforming design for wireless information and power transfer systems: receive power-splitting versus transmit time-switching'. *IEEE Transactions on Communications*. 2017, vol. 65(2), pp. 876–89.

[270]   Holis J., Pechac P. 'Elevation dependent shadowing model for mobile communications via high altitude platforms in built-up areas'. *IEEE Transactions on Antennas and Propagation*. 2008, vol. 56(4), pp. 1078–84.

[271]   Mozaffari M., Saad W., Bennis M., Debbah M. 'Unmanned aerial vehicle with underlaid device-to-device communications: performance and trade-offs'. *IEEE Transactions on Wireless Communications*. 2016, vol. 15(6), pp. 3949–63.

[272]   Nguyen L.D., Nguyen K.K., Kortun A., Duong T.Q. 'Realtime deployment and resource allocation for distributed UAV systems in disaster relief'. IEEE SPAWC, Cannes, France; 2019. p. 1.

[273]   Mase K. 'How to deliver your message from/to a disaster area'. *IEEE Communications Magazine*. 2011, vol. 49(1), pp. 52–7.

[274]   Masaracchia A., Nguyen L.D., Duong T.Q., Nguyen N.M. 'An energy-efficient clustering and routing framework for disaster relief network'. *IEEE Access: Practical Innovations, Open Solutions*. 2019.1.

[275]   Duong T.Q., Nguyen L.D., Nguyen L.K. 'Practical optimisation of path planning and completion time of data collection for UAV-enabled disaster communications'. IEEE IWCMC, Tangier, Morocco; 2019. pp. 1–5.

[276]   Zeng Y., Zhang R., Lim T.J. 'Throughput maximization for UAV-enabled mobile relaying systems'. *IEEE Transactions on Communications*. 2016, vol. 64(12), pp. 4983–96.

[277]   Nasir A.A., Tuan H.D., Duong T.Q., Poor H.V. 'UAV-enabled communication using NOMA'. *IEEE Transactions on Communications*. 2019, vol. 67(7), pp. 5126–38.

[278]   Wagstaff K., Cardie C., Rogers S., Schrödl S. 'Constrained K-means clustering with background knowledge'. Proceedings of the Eighteenth International Conference on Machine Learning; 2001. pp. 577–84.

[279]   Marzetta T.L., Larsson E.G., Yang H., Ngo H.Q. *Fundamentals of massive MIMO*. Cambridge University Press; 2016.

[280]   Sung C.K., Collings I.B. 'Multiuser cooperative multiplexing with interference suppression in wireless relay networks'. *IEEE Transactions on Wireless Communications*. 2010, vol. 9(8), pp. 2528–38.

[281]   Alzenad M., El-Keyi A., Yanikomeroglu H. '3-D placement of an unmanned aerial vehicle base station for maximum coverage of users with different QoS requirements'. *IEEE Wireless Communications Letters*. 2018, vol. 7(1), pp. 38–41.

[282]   Zhan C., Zeng Y., Zhang R. 'Energy-efficient data collection in UAV enabled wireless sensor network'. *IEEE Wireless Communications Letters*. 2018, vol. 7(3), pp. 328–31.

[283]   Park H.-S., Jun C.-H. 'A simple and fast algorithm for k-medoids clustering'. *Expert Systems with Applications*. 2009, vol. 36(2), pp. 3336–41.

[284]   Zeng Y., Xu X., Zhang R. 'Trajectory design for completion time minimization in UAV-enabled multicasting'. *IEEE Transactions on Wireless Communications*. 2018, vol. 17(4), pp. 2233–46.

[285]   Khawaja W., Guvenc I., Matolak D., Fiebig U.-C., Schneckenberger N. 'A survey of air-to-ground propagation channel modeling for unmanned aerial vehicles'. *arXiv preprint arXiv:1801.01656*. 2018.

[286]   Lethaby N. 'Wireless connectivity for the internet of things: one size does not fit all'. Texas Instruments; 2017.

[287]   Duong T.Q., Nguyen L.D., Tuan H.D., Hanzo L. 'Learning-aided realtime performance optimisation of cognitive UAV-assisted disaster communication'. IEEE GLOBECOM, Hawaii, US, Dec; 2019. p. 1.

[288]   Liu X., Guan M., Zhang X., Ding H. 'Spectrum sensing optimization in an UAV-based cognitive radio'. *IEEE Access*. 2018, vol. 6, pp. 44002–9.

[289]   Sboui L., Ghazzai H., Rezki Z., Alouini M.-S. 'Achievable rates of UAV-relayed cooperative cognitive radio MIMO systems'. *IEEE Access*. 2017, vol. 5, pp. 5190–204.

[290]   L.Wang., Yang H., Long J., K.Wu., Chen J. 'Enabling ultra-dense UAV-aided network with overlapped spectrum sharing: potential and approaches'. *IEEE Network*. 2018, vol. 32(5), pp. 85–91.

[291]   Huang Y., Xu J., Qiu L., Zhang R. 'Cognitive UAV communication via joint trajectory and power control'. The 9th IEEE International Workshop on Signal Processing Advances in Wireless Communications (SPAWC); 2008. pp. 1–5.

[292]   Sboui L., Ghazzai H., Rezki Z., Alouini M. 'Energy-efficient power allocation for UAV cognitive radio systems'. IEEE 86th Vehicular Technology Conference: VTC2017-Fall; Toronto, Canada; 2017. pp. 1–5.

[293]   Sun H., Chen X., Shi Q., Hong M., Fu X., Sidiropoulos N.D. 'Learning to optimize: training deep neural networks for interference management'. *IEEE Transactions on Signal Processing*. 2018, vol. 66(20), pp. 5438–53.

[294] Nguyen L.D., Tuan H.D., Duong T.Q., Poor H.V. 'Multi-user regularized zero-forcing beamforming'. IEEE Transactions on Signal Processing : A Publication of the IEEE Signal Processing Society; 2019. p. 1.

[295] Tuy H. *Convex analysis and global optimization.* second Edition. Springer; 2016.

Appendices

A.1 Appendix A

This section provides some useful matrix analysis for the approximation of complexity functions.

A.1.1 Basic vector and matrix calculation

Product of vector and matrix

$< \mathbf{x}, \mathbf{y} > = \mathbf{x}^H \mathbf{y} = \sum_{i=1}^{N} x_i^H y_i$

for $\mathbf{x}, \mathbf{y} \in \mathbb{C}^N$,

$< \mathbf{X}, \mathbf{Y} > = Tr(\mathbf{X}^H \mathbf{Y}) = \sum_{i=1, j=1}^{M,N} X_{i,j}^H Y_{i,j}$

for $\mathbf{X}, \mathbf{Y} \in \mathbb{C}^{M \times N}$.

The definition of unitary matrix

$\mathbf{Q}\mathbf{Q}^H = \mathbf{Q}^H\mathbf{Q} = I$

Kronecker product:

For $\mathbf{A} \in \mathbb{R}^{M \times N}$ and $\mathbf{B} \in \mathbb{R}^{P \times Q}$, the Kronecker product of $\mathbf{A}$ and $\mathbf{B}$, $\mathbf{A} \otimes \mathbf{B} \in \mathbb{R}^{MP \times NQ}$, is defined as

$$\mathbf{A} \otimes \mathbf{B} = \begin{bmatrix} A_{11}\mathbf{B} & ... & A_{1N}\mathbf{B} \\ ... & ... & ... \\ A_{M1}\mathbf{B} & ... & A_{MN}\mathbf{B} \end{bmatrix}$$

A.1.2 Matrix Norm

Amplification factor (gain) of $\mathbf{A}$

$$\|\mathbf{A}\|_2 = \max_{\mathbf{X} \neq 0} \frac{\|\mathbf{AX}\|}{\|\mathbf{X}\|} = \sqrt{\lambda_{max}(\mathbf{A}^T\mathbf{A})} \tag{A.1}$$

$$\min_{\mathbf{X} \neq 0} \frac{\|\mathbf{AX}\|}{\|\mathbf{X}\|} = \sqrt{\lambda_{min}(\mathbf{A}^T\mathbf{A})} \tag{A.2}$$

A.1.3   Logarithm of positive definite matrices

For $\mathbf{A} \in \mathbb{R}^{n \times n}$ is a positive definite :

$$\log(A) = Q \begin{bmatrix} \log(\lambda_1) & & \\ & \ldots & \\ & & \log(\lambda_d) \end{bmatrix} \mathbf{Q}^H$$

where

$$A = Q \begin{bmatrix} \lambda_1 & & \\ & \ldots & \\ & & \lambda_d \end{bmatrix} \mathbf{Q}^H$$

$\lambda_1, \ldots, \lambda_n$ are positive eigenvalues of $\mathbf{A}$ and $\mathbf{Q}$ is unitary.

Logarithm of Kronecker product:

$$\log(\mathbf{A} \otimes \mathbf{H}) = \log(\mathbf{A}) \otimes \mathbf{I} + \mathbf{I} \otimes \log(\mathbf{H})$$

where $\mathbf{A}, \mathbf{H}$ are positive definite matrices and $\mathbf{I}$ is an identity matrix.

A.1.4   Trace and logarithm relationship

For a symmetric positive semidefinite (PSD) matrix $\mathbf{A}$

$$\ln(\det(\mathbf{A})) = \mathrm{Tr}(\ln(\mathbf{A})) = \sum_i^n \ln(\lambda_i)$$

where λ_i is the eigenvalue of $\mathbf{A}$.

A.1.5   Some case studies of complex matrices

- Case 1: For any matrix $\mathbf{A} \in \mathbb{C}^{M \times K}$

$$(\mathbf{A}\mathbf{A}^H + \mathbf{I}_M)^{-1} = \mathbf{I}_M - \mathbf{A}(\mathbf{A}^H\mathbf{A} + \mathbf{I}_K)^{-1}\mathbf{A}^H$$

- Case 2: For square matrices $\mathbf{A}, \mathbf{B}$:

$$(\mathbf{A} + \delta\mathbf{B}) = (\mathbf{A}\mathbf{B}^{-1}/\delta + I)\delta\mathbf{B}$$
$$(\mathbf{A} + \delta\mathbf{B})^{-1} = \frac{1}{\delta}\mathbf{B}^{-1}(\frac{1}{\delta}\mathbf{A}\mathbf{B}^{-1} + \mathbf{I})^{-1}$$

where $\delta \in \mathbb{R}_{++}$. Assuming $|\det(\frac{1}{\delta}\mathbf{A}\mathbf{B}^{-1})| < 1$

$$(\mathbf{A} + \delta\mathbf{B})^{-1} = -\sum_{n \geq 0}(-\delta)^{-(n+1)}\mathbf{B}^{-1}(\mathbf{A}\mathbf{B}^{-1})^n$$

- Case 3: For square matrices $\mathbf{M}, \mathbf{D} = d\mathbf{I}$:

$$(\mathbf{M} + \mathbf{D})^{-1} = \mathbf{D}^{-1}(\mathbf{M}\mathbf{D}^{-1} + \mathbf{I})^{-1}$$

Let $\mathbf{M}\mathbf{D}^{-1} = \mathbf{Q}\boldsymbol{\Lambda}\mathbf{Q}^H$, then

$$\mathbf{D}^{-1}(\mathbf{Q}\boldsymbol{\Lambda}\mathbf{Q}^H + \mathbf{I})^{-1} = \mathbf{D}^{-1}[\mathbf{Q}(\boldsymbol{\Lambda} + \mathbf{I})\mathbf{Q}^H]^{-1} = \mathbf{D}^{-1}\mathbf{Q}(\boldsymbol{\Lambda} + \mathbf{I})^{-1}\mathbf{Q}^H$$

where $D^{-1} = \mathrm{diag}[\frac{1}{D_{ii}}]$

A.1.6 Schur complement

$$\mathbf{A} = \begin{bmatrix} \mathbf{A}_{11} & \mathbf{A}_{12} \\ \mathbf{A}_{21} & \mathbf{A}_{22} \end{bmatrix}, \quad \mathbf{A}_{11} : \text{non-singular}$$

$$\mathbf{A}\mathbf{X} = 0 \Leftrightarrow (\mathbf{A}_{22} - \mathbf{A}_{21}\mathbf{A}_{11}^{-1}\mathbf{A}_{12})x = 0$$

Hence, $\mathbf{A}/\mathbf{A}_{11} = \mathbf{A}_{22} - \mathbf{A}_{21}\mathbf{A}_{11}^{-1}\mathbf{A}_{12}$ (Schur complement of $\mathbf{A}_{11}$ in $\mathbf{A}$)

A.1.7 Matrix Inverse Analysis

$\mathbf{A}$ is a square matrix, $\mathbf{A}$ is said to be nonsingular if $\mathbf{A}^{-1}$ exists:

$$\mathbf{A}\mathbf{A}^{-1} = \mathbf{A}^{-1}\mathbf{A} = \mathbf{I}$$

If $\mathbf{A}$ is singular then $\mathbf{A}^{-1}$ does not exist and
  If $\mathbf{A}$ is singular then does not exist and

$$\det(\mathbf{A}) = 0$$

For $\mathbf{M} = \mathbf{A} + i\mathbf{B}$, $\mathbf{Z} = \mathbf{X} + iy$, $\mathbf{M}, \mathbf{Z} \in \mathbb{C}^{N \times N}$ where $\mathbf{A}, \mathbf{B}, \mathbf{X}, \mathbf{Y}$ are real matrices:

$$(\mathbf{A} + i\mathbf{B})^{-1} = (\mathbf{A} + \mathbf{B}\mathbf{A}^{-1}\mathbf{B})^{-1} - i(\mathbf{B} + \mathbf{A}\mathbf{B}^{-1}\mathbf{A})^{-1}$$

Proof. Suppose

$$(\mathbf{A} + i\mathbf{B})(\mathbf{X} + i\mathbf{Y}) = (\mathbf{A}\mathbf{X} - \mathbf{B}\mathbf{Y}) + i(\mathbf{B}\mathbf{X} + \mathbf{A}\mathbf{Y}) = \mathbf{I}$$

$$\begin{cases} \mathbf{A}\mathbf{X} - \mathbf{B}\mathbf{Y} = \mathbf{I} & \Rightarrow \mathbf{A}\mathbf{X} - \mathbf{B}\mathbf{Y} + \mathbf{B}\mathbf{A}^{-1}(\mathbf{B}\mathbf{X} + \mathbf{A}\mathbf{Y}) = \mathbf{I} & \Rightarrow \mathbf{X} = (\mathbf{A} + \mathbf{B}\mathbf{A}^{-1}\mathbf{B})^{-1} \\ \mathbf{B}\mathbf{X} + \mathbf{A}\mathbf{Y} = 0 & \Rightarrow \mathbf{A}\mathbf{X} - \mathbf{B}\mathbf{Y} - \mathbf{A}\mathbf{B}^{-1}(\mathbf{B}\mathbf{X} + \mathbf{A}\mathbf{Y}) = \mathbf{I} & \Rightarrow \mathbf{Y} = -(\mathbf{B} + \mathbf{A}\mathbf{B}^{-1}\mathbf{A})^{-1} \end{cases}$$

where $\mathbf{0}$ is zero matrix.

Sherman Morrison Lemma

Whenever $(\mathbf{A} + \mathbf{B})^{-1}$ exists, $\mathbf{B}$ has rank $1, g = \text{Tr}(\mathbf{BA}^{-1})$, and $g \neq 1$:

$$(\mathbf{A} + \mathbf{B})^{-1} = \mathbf{A}^{-1} - \frac{1}{1+g}\mathbf{A}^{-1}\mathbf{BA}^{-1}$$

$$(\mathbf{A} + \mathbf{uv})^{-1} = \mathbf{A}^{-1} - \frac{(\mathbf{A}^{-1}\mathbf{u})(\mathbf{v}^T\mathbf{A}^{-1})}{1+\mathbf{v}^T\mathbf{A}^{-1}\mathbf{u}}$$

$$\mathbf{x}^H(\mathbf{A} + c\mathbf{xx}^H)^{-1} = \frac{\mathbf{x}^H\mathbf{A}^{-1}}{1+c\mathbf{x}^H\mathbf{A}^{-1}\mathbf{x}}$$

Woodbury matrix identity:

$$(\mathbf{A} + \mathbf{UCV})^{-1} = \mathbf{A}^{-1} - \mathbf{A}^{-1}\mathbf{U}(\mathbf{C}^{-1} + \mathbf{VA}^{-1}\mathbf{U})^{-1}\mathbf{VA}^{-1}$$

Let $\mathbf{U} = \mathbf{V} = \mathbf{I}$, then

$$(\mathbf{A} + \mathbf{C})^{-1} = \mathbf{A}^{-1} - \mathbf{A}^{-1}(\mathbf{C}^{-1} + \mathbf{A}^{-1})^{-1}\mathbf{A}^{-1}$$

Sherman-Morrison-Woodbury

$$(\beta_1\mathbf{I} + \beta_2\mathbf{A}^T\mathbf{A})^{-1} = \frac{1}{\beta_1}\mathbf{I} - \frac{\beta_2}{\beta_1}\mathbf{A}^T(\beta_1\mathbf{I} + \beta_2\mathbf{AA}^T)^{-1}\mathbf{A}$$

for all $\beta_1 > 0, \beta_2 > 0$.

Jensen's inequality

$$(1 - \lambda)\mathbf{X}^{-1} + \lambda\mathbf{Y}^{-1} \succeq ((1 - \lambda)\mathbf{X} + \lambda\mathbf{Y})^{-1}$$

where $\mathbf{X}$ and $\mathbf{Y}$ are positive definite matrices.

Cauchy-Schwarz inequality:

For variable vectors:

$$|\mathbf{x}^H\mathbf{y}|^2 = \mathbf{x}^H\mathbf{yy}^H\mathbf{x} \leq \mathbf{x}^H\mathbf{xy}^H\mathbf{y}$$

For variable matrices,

$$\mathbf{X}^H\mathbf{Y}(\mathbf{Y}^H\mathbf{Y})^{-1}\mathbf{Y}^H\mathbf{X} \leq \mathbf{X}^H\mathbf{Y}$$

A.1.8   *Sum of matrices inversion*

$$\mathbf{A} + \mathbf{B} = \mathbf{B}(\mathbf{B}^{-1} + \mathbf{B}^{-1}\mathbf{AB}^{-1})\mathbf{B}$$

$$\mathbf{A} + \mathbf{I} = (\mathbf{A}^{-1} + \mathbf{I})\mathbf{A} \Leftrightarrow \mathbf{A}^{-1} + \mathbf{I} = (\mathbf{A} + \mathbf{I})\mathbf{A}^{-1}$$

$$\mathbf{A}^{-1} + \mathbf{B}^{-1} = \mathbf{B}^{-1}(\mathbf{B} + \mathbf{BA}^{-1}\mathbf{B})\mathbf{B}^{-1}$$

$$(\mathbf{A}^{-1} + \mathbf{I})^{-1} = \mathbf{A}(\mathbf{A} + \mathbf{I})^{-1} = \mathbf{I} - (\mathbf{A} + \mathbf{I})^{-1}$$
$$(\mathbf{A}^{-1} + \mathbf{B}^{-1})^{-1} = \mathbf{A}(\mathbf{A} + \mathbf{B})^{-1}\mathbf{B} = \mathbf{A} - \mathbf{A}(\mathbf{A} + \mathbf{B})^{-1}\mathbf{A} = \mathbf{B} - \mathbf{B}(\mathbf{A} + \mathbf{B})^{-1}\mathbf{B}$$

A.1.9   *Inversion Identities*

$$(\mathbf{A}^{-1} + \mathbf{U}\mathbf{V}^H)^{-1} = \mathbf{A} - \mathbf{A}\mathbf{U}(\mathbf{I} + \mathbf{V}^H\mathbf{A}\mathbf{U})^{-1}\mathbf{V}^H\mathbf{A}$$

$$(\mathbf{A} + \mathbf{U}\mathbf{V}^H)^{-1} = \mathbf{A}^{-1} - \mathbf{A}^{-1}\mathbf{U}(\mathbf{I} + \mathbf{V}^H\mathbf{A}^{-1}\mathbf{U})^{-1}\mathbf{V}^H\mathbf{A}^{-1}$$

$$(\mathbf{A}^{-1} + \mathbf{u}\mathbf{v}^H)^{-1} = \mathbf{A} - \frac{\mathbf{A}\mathbf{u}\mathbf{v}^H\mathbf{A}}{1 + \mathbf{v}^H\mathbf{A}\mathbf{u}}$$

$$(\mathbf{A}^{-1} + \mathbf{U}\mathbf{C}^{-1}\mathbf{V}^H)^{-1} = \mathbf{A} - \mathbf{A}\mathbf{U}(\mathbf{C} + \mathbf{V}^H\mathbf{A}\mathbf{U})^{-1}\mathbf{V}^H\mathbf{A}$$

$$(\mathbf{A}^{-1} + \mathbf{V}\mathbf{V}^H)^{-1} = \mathbf{A} - \mathbf{A}\mathbf{V}(\mathbf{I} + \mathbf{V}^H\mathbf{A}\mathbf{V})^{-1}\mathbf{V}^H\mathbf{A}$$

$$(\mathbf{A}^{-1} + \mathbf{U}\mathbf{V}^H)^{-1}\mathbf{U} = \mathbf{A}\mathbf{U}(\mathbf{I} + \mathbf{V}^H\mathbf{A}\mathbf{U})^{-1}$$

$$\mathbf{V}^H(\mathbf{A}^{-1} + \mathbf{U}\mathbf{V}^H)^{-1} = (\mathbf{I} + \mathbf{V}^H\mathbf{A}\mathbf{U})^{-1}\mathbf{V}^H\mathbf{A}$$

$$\mathbf{V}^H(\mathbf{A}^{-1} + \mathbf{U}\mathbf{V}^H)^{-1}\mathbf{U} = \mathbf{I} - (\mathbf{I} + \mathbf{V}^H\mathbf{A}\mathbf{U})^{-1}$$

$$(\mathbf{A}^{-1} + \mathbf{V}\mathbf{V}^H)^{-1} = \mathbf{A} - \mathbf{A}\mathbf{V}(\mathbf{I} + \mathbf{V}^H\mathbf{A}\mathbf{V})^{-1}\mathbf{V}^H\mathbf{A}$$

A.1.10   *Determination inversion*

$$\det(\mathbf{A}^{-1} + \mathbf{U}\mathbf{V}^H) = \det(1 + \mathbf{V}^H\mathbf{A}\mathbf{U})\det(\mathbf{A}^{-1})$$

$$\det(\mathbf{A}^{-1} + \mathbf{u}\mathbf{v}^H) = (1 + \mathbf{v}^H\mathbf{A}\mathbf{u})\det(\mathbf{A}^{-1})$$

A.1.11   *Log-determinant of a matrix (*log det(.)*)*

Whenever $\mathbf{X} \succ 0$:

$$\log\det(\mathbf{X}^{-1}) = \log\det(\mathbf{X})^{-1}$$
$$= -\log\det(\mathbf{X}) = \log\det(\mathrm{diag}[1/\lambda_i]_i)$$
$$= \log(\textstyle\prod_i 1/\lambda_i) = \textstyle\sum_i \log(1/\lambda_i)$$

where λ_i is the ith eigenvalue of $\mathbf{X}$.

Whenever $\mathbf{A}, \mathbf{B}$ are non-singular:

$$\log\det(\mathbf{I} + \mathbf{A}\mathbf{B}^{-1}) = \log\det(\mathbf{A}(\mathbf{A}^{-1} + \mathbf{B}^{-1}))$$
$$= \log(\det(\mathbf{A})\det(\mathbf{A}^{-1} + \mathbf{B}^{-1}))$$
$$= \log\det(\mathbf{A}) + \log\det(\mathbf{A}^{-1} + \mathbf{B}^{-1}).$$

A.2 Appendix B

Some examples of equality and inequality for approximation of non-convex functions are provided as below.

A.2.1 Some equalities and inequalities in $\mathbb{R}$

- Inequality 1

$$x^2 \geq 2\bar{x}x - \bar{x}^2, \tag{A.3}$$

for all $x > 0, \bar{x} > 0$.

- Inequality 2:

$$\frac{x}{t} = \left(\frac{\sqrt{x}}{\sqrt{t}}\right)^2 \geq \frac{2\sqrt{\bar{x}}\sqrt{x}}{\bar{t}} - \frac{\bar{x}}{\bar{t}^2}t \tag{A.4}$$

for all $x > 0, \bar{x} > 0, t > 0, \bar{t} > 0$. Especially,

$$\frac{1}{t} \geq \frac{2}{\bar{t}} - \frac{t}{\bar{t}^2} \tag{A.5}$$

for all $t > 0, \bar{t} > 0$.

- Inequality 3

$$|a+b| - |b| \geq -|a| \tag{A.6}$$

- Inequality 4

$$\frac{1}{x+a} \geq \frac{a + 2\bar{x} - x}{(\bar{x}+a)^2} \tag{A.7}$$

for all $x + a \neq 0$.

- Inequality 5

$$\|x+y\| \leq \|x\| + \|y\| \tag{A.8}$$

$$\|Ax\| \leq \|A\| . \|x\| \tag{A.9}$$

where x, y are vectors and A is a matrix.

- Inequality 6 (Holder inequality)

$$|\langle x, y \rangle| = |x^H y| \le \|x\|_p + \|y\|_q : \frac{1}{p} + \frac{1}{q} = 1 \tag{A.10}$$

- Inequality 7

$$\|x\|_1 \le \sqrt{n}\|x\|_2, \quad \|x\|_1 \le n\|x\|_\infty \tag{A.11}$$

$$\|x\|_2 \le \|x\|_1, \quad \|x\|_2 \le \sqrt{n}\|x\|_\infty \tag{A.12}$$

$$\|x\|_\infty \le \|x\|_1, \quad \|x\|_\infty \le \|x\|_2 \tag{A.13}$$

A.2.2   Some norm inequalities of square matrix in $\mathbb{R}^{n \times n}$

- $\|A\|_1 \le \sqrt{n}\|A\|_2$ $\|A\|_1 \le n\|A\|_\infty$ $\|A\|_1 \le \sqrt{n}\|A\|_F$
- $\|A\|_2 \le \sqrt{n}\|A\|_1$ $\|A\|_2 \le \sqrt{n}\|A\|_\infty$ $\|A\|_2 \le \|A\|_F$
- $\|A\|_\infty \le n\|A\|_1$ $\|A\|_\infty \le \sqrt{n}\|A\|_2$ $\|A\|_\infty \le \sqrt{n}\|A\|_F$
- $\|A\|_F \le \sqrt{n}\|A\|_1$ $\|A\|_F \le \sqrt{n}\|A\|_2$ $\|A\|_F \le \sqrt{n}\|A\|_\infty$
- $\max_{i,j} |a_{i,j}| \le \|A\|_2 \le \sqrt{mn} \max_{i,j} |a_{i,j}|$

A.3   Appendix C

Some examples of first-order approximations are shown as below. These approximations can be applied for proposing convex optimisation problems in many applications of signal processing and wireless communication [25, 51, 147–150, 259].

A.3.1   Some inequalities using first-order approximation for determining lower bound of complex functions

- As the function $f(z, t) = 1/zt$ is convex in $z > 0, t > 0$, we have the following inequality

$$\frac{1}{zt}(\bar{z}, \bar{t}) + \langle \nabla f(\bar{z}, \bar{t}), (z, t) - (\bar{z}, \bar{t}) \rangle = 3\frac{1}{\bar{z}\bar{t}} - \left(\frac{z/\bar{z} + t/\bar{t}}{\bar{z}\bar{t}} \right), \tag{A.14}$$

for all $z > 0,\ \bar{z} > 0, t > 0, \bar{t} > 0$.

- For all $\alpha, \beta \geq 0$, $\alpha + \beta = 1$, $\mathbf{V}_1, \mathbf{V}_2, \mathbf{X}_1, \mathbf{X}_2 \succeq 0$

$$(\alpha\mathbf{V}_1 + \beta\mathbf{V}_2)^H(\alpha\mathbf{X}_1 + \beta\mathbf{X}_2)^{-1}(\alpha\mathbf{V}_1 + \beta\mathbf{V}_2) \preceq \alpha\mathbf{V}_1^H\mathbf{X}_1^{-1}\mathbf{V}_1 + \beta\mathbf{V}_2^H\mathbf{X}_2^{-1}\mathbf{V}_2$$

A.3.2 Some inequalities using first-order approximation for determining upper bound of complex functions

- For all $z > 0$, $\bar{z} > 0, t > 0, \bar{t} > 0$, and $2z - \bar{z} > 0$, $2t - \bar{t} > 0$:

$$\frac{1}{zt} \leq \frac{1}{2}\left(\frac{\bar{z}}{\bar{t}}\frac{1}{z^2} + \frac{\bar{t}}{\bar{z}}\frac{1}{t^2}\right) \leq \frac{1}{2}\left(\frac{\bar{z}}{\bar{t}}\frac{1}{2\bar{z}z - \bar{z}^2} + \frac{\bar{t}}{\bar{z}}\frac{1}{2\bar{t}t - \bar{t}^2}\right) \tag{A.15}$$

- For all $x > 0$, $\bar{x} > 0, y > 0, \bar{y} > 0, z > 0, \bar{z} > 0$, and $\sqrt{t} \geq 1/yz$, $2x - \bar{x} > 0$:

$$\frac{1}{xyz} \leq \frac{1}{2}\left(\frac{\bar{x}}{\bar{y}\bar{z}}\frac{1}{x^2} + \frac{\bar{y}\bar{z}}{\bar{x}}\frac{1}{y^2z^2}\right) \leq \frac{1}{2}\left(\frac{\bar{x}}{\bar{y}\bar{z}}\frac{1}{2\bar{x}x - \bar{x}^2} + \frac{\bar{y}\bar{z}}{\bar{x}}t\right) \tag{A.16}$$

- For all $z > 0$, $\bar{z} > 0, t > 0, \bar{t} > 0$:

$$zt \leq \frac{1}{2}\left[\frac{\bar{t}}{\bar{z}}(2\bar{z}z - \bar{z}^2) + \frac{\bar{z}}{\bar{t}}(2\bar{t}t - \bar{t}^2)\right] \leq \frac{1}{2}\left[\frac{\bar{t}}{\bar{z}}z^2 + \frac{\bar{z}}{\bar{t}}t^2\right] \tag{A.17}$$

- For all $x > 0$, $\bar{x} > 0, y > 0, \bar{y} > 0, z > 0, \bar{z} > 0$:

$$xyz \leq \frac{1}{2}\left[\frac{\bar{y}\bar{z}}{\bar{x}}x^2 + \frac{\bar{x}}{\bar{y}\bar{z}}(yz)^2\right] \leq \frac{1}{2}\left[\frac{\bar{y}\bar{z}}{\bar{x}}x^2 + \frac{\bar{x}}{\bar{y}\bar{z}}\frac{1}{4}\left(\frac{\bar{z}}{\bar{y}}y^2 + \frac{\bar{y}}{\bar{z}}z^2\right)^2\right] \tag{A.18}$$

- As the function

$$f(\mathbf{S}, \mathbf{Y}) = (\mathbf{S}^H\mathbf{Y}^{-1}\mathbf{S})$$

is a convex in $\mathbf{Y} > 0$ and $\mathbf{S}$. We have found the following inequality

$$f(\mathbf{S}, y) \geq f(\bar{\mathbf{S}}, \bar{\mathbf{Y}}) + \langle \nabla f(\bar{\mathbf{S}}, \bar{\mathbf{Y}}), (\mathbf{S}, \mathbf{Y}) - (\bar{\mathbf{S}}, \bar{\mathbf{Y}})\rangle$$

$$= 2\text{Re}\{(\bar{\mathbf{S}}^H\bar{\mathbf{Y}}^{-1}\mathbf{S})\} - \text{Tr}(\bar{\mathbf{S}}^H\bar{\mathbf{Y}}^{-1}\mathbf{Y}\bar{\mathbf{Y}}^{-1}\bar{\mathbf{S}}) \tag{A.19}$$

A.3.3 Some logarithm inequalities using first-order approximations

- For all $z > 0$, $\bar{z} > 0$:

$$\ln(1 + z) \geq \ln(1 + \bar{z}) + \frac{\bar{z}}{\bar{z} + 1} - \frac{\bar{z}^2}{\bar{z} + 1}\frac{1}{z}(\text{A.20})$$

Proof:

Note that function $f(x) - \ln(1 + 1/x)$ is convex in $x > o$. For any and x>0 and $\bar{x}$ >0, it is true that [295]

$$\ln\left(1 + \frac{1}{x}\right) \geq \ln\left(1 + \frac{1}{\bar{x}}\right) + \left(\frac{\nabla f(\bar{x})}{\nabla x}\right)(x - \bar{x}) = \ln\left(1 + \frac{1}{\bar{x}}\right) + \frac{1}{1+\bar{x}} - \frac{1}{(1+\bar{x})\bar{x}}x.$$

$$(\text{A.21})$$

Then (A.19) is obtained by replacing $z = 1/x$ and $\bar{z} = 1/\bar{x}$ into (A.20)

- For all $t > 0, \bar{t} > 0$, $z > 0, \bar{z} > 0$:

$$t \ln(1 + z) \geq a - \frac{b}{z} - \frac{c}{t}$$

$$(\text{A.22})$$

where

$$a = 2\bar{t}\ln(1 + \bar{x}z) + \frac{\bar{t}\bar{z}}{\bar{z}+1} > 0, b = \frac{\bar{t}\bar{z}^2}{\bar{z}+1} > 0, c = \bar{t}^2\ln(1 + \bar{z}) > 0.$$

Proof:

The function $f(x, t) = \ln(1 + 1/x)/t$ is convex in $x > 0, t > 0$ which can be proved by examining its Hessian. The following inequality for all $x > 0$, $\bar{x} > 0$, $t > 0$ and $\bar{t} > 0$ then holds true:

$$\frac{\ln(1 + 1/x)}{t} \geq f(\bar{x}, \bar{t}) + \langle \nabla f(\bar{x}, \bar{t}), (x, t) - (\bar{x}, \bar{t}) \rangle = 2\frac{\ln(1 + 1/\bar{x})}{\bar{t}} + \frac{1}{\bar{t}(\bar{x} + 1)} - \frac{x}{(\bar{x} + 1)\bar{x}\bar{t}} - \frac{\ln(1 + 1/\bar{x})}{\bar{t}^2}t.$$

$$(\text{A.23})$$

By replacing $1/x \to z$, $1/\bar{x} \to \bar{z}$, $1/t \to t$, and $1/\bar{t} \to \bar{t}$ in (A.23) we found (A.21)

- Using (A.22) and replacing $1/x \to x$ and $1/\bar{x} \to \bar{x}$, we have

$$\frac{\ln(1+x)}{t} \geq a - \frac{b}{x} - ct,$$

$$(\text{A.24})$$

where

$$a = 2\frac{\ln(1+\bar{x})}{\bar{t}} + \frac{\bar{x}}{\bar{t}(\bar{x}+1)} > 0, b = \frac{\bar{x}^2}{\bar{t}(\bar{x}+1)} > 0, c = \frac{\ln(1+\bar{x})}{\bar{t}^2} > 0.$$

Using (A.22) and replacing $|x|^2 \to x$ and $|\bar{x}|^2 \to \bar{x}$ we have

$$\frac{\ln(1+|x|^2)}{t} \geq \bar{a} - \frac{\bar{b}}{|x|^2} - \bar{c}t \geq \bar{a} - \frac{\bar{b}}{2\Re\{x\bar{x}^*\} - |\bar{x}|^2} - \bar{c}t \qquad (A.25)$$

over the trust region

$$2\Re\{x\bar{x}^*\} - |\bar{x}|^2 > 0, \qquad (A.26)$$

where

$$\bar{a} = 2\frac{\ln(1+|\bar{x}|^2)}{\bar{t}} + \frac{|\bar{x}|^2}{\bar{t}(|\bar{x}|^2+1)} > 0, \bar{b} = \frac{|\bar{x}|^4}{\bar{t}(|\bar{x}|^2+1)} > 0, \bar{c} = \frac{\ln(1+|\bar{x}|^2)}{\bar{t}^2} > 0.$$

- We are following inequalities

$$\ln(1 + \frac{|x|^2}{y}) \geq \ln(1 + \frac{|\bar{x}|^2}{\bar{y}}) - \frac{|\bar{x}|^2}{\bar{y}} + 2\frac{\Re\{\bar{x}^*x\}}{\bar{y}} - \frac{|\bar{x}|^2(|x|^2 + y)}{\bar{y}(\bar{y} + |\bar{x}|^2)}, \qquad (A.27)$$

$$\frac{|x|^2}{y} \geq 2\frac{\bar{x}^*x}{\bar{y}} - \frac{|\bar{x}|^2}{\bar{y}^2}y, \qquad (A.28)$$

$$\forall x \in \mathbb{C}, \bar{x} \in \mathbb{C}, y > 0, \bar{y} > 0.$$

Proof:

Function $f(t) = -\ln(1 - t)$ is obviously convex and increasing in the domain $0 \leq t < 1$, while function $g(x, z) = |x|^2/z$ is convex. Therefore, the composite function $f(g(x, y)) = -\ln(1 - |x|^2/z)$ is convex in the domain $z > |x|^2$[295], for which

$$-\ln\left(1 - \frac{|x|^2}{z}\right) \geq -\ln\left(1 - \frac{|\bar{x}|^2}{\bar{z}}\right) + \langle\nabla f(g(\bar{x}, \bar{z})), (x, z) - (\bar{x}, \bar{z})\rangle = -\ln\left(1 - \frac{|\bar{x}|^2}{\bar{z}}\right) - \frac{|\bar{x}|^2}{\bar{z}-|\bar{x}|^2} +$$
$$2\frac{\Re\{\bar{x}^*x\}}{\bar{z}-|\bar{x}|^2} - \frac{|\bar{x}|^2z}{(\bar{y}-|\bar{x}|^2\bar{z})}.$$

$$(A.29)$$

By noting

$$\ln\left(1 + \frac{|x|^2}{y}\right) = -\ln\left(1 - \frac{|x|^2}{y+|x|^2}\right) \qquad (A.30)$$

(A.27) is obtained by applying (A.29) for $z = y + |x|^2$ and $\bar{z} = \bar{y} + |\bar{x}|^2$.

Furthermore, as $g(x, y) = |x|^2/y$ is convex in $x \in \mathbb{C}$ and $y > 0$, it is true that [295]

$$\frac{|x|^2}{y} \geq \frac{|\bar{x}|^2}{\bar{y}} + \langle\nabla g(\bar{x}, \bar{y}), (x, y) - (\bar{x}, \bar{y})\rangle = 2\frac{\bar{x}^*x}{\bar{y}} - \frac{|\bar{x}|^2}{\bar{y}^2}y \qquad (A.31)$$

showing (A.27) and the proof is completed.

Note that function $f(x, y, t) = \frac{\ln(1 + 1/xy)}{t}$ is convex in $x > 0, y > 0, t > 0$ which can be proved by examining its Hessian. The following inequality for all $x > 0, \bar{x} > 0, y > 0, \bar{y} > 0, t > 0$ and $\bar{t} > 0$ then holds true [295]:

$$\frac{\ln(1 + 1/xy)}{t} \geq f(\bar{x}, \bar{y}, \bar{t}) + \langle\nabla f(\bar{x}, \bar{y}, \bar{t}), (x, y, t) - (\bar{x}, \bar{y}, \bar{t})\rangle = a - bx - cy - dt,$$

where

$$a = 2\frac{\ln(1 + 1/\bar{x}\bar{y})}{\bar{t}} + \frac{2}{\bar{t}(\bar{x}\bar{y} + 1)} > 0, b = \frac{1}{(\bar{x}\bar{y} + 1)\bar{x}\bar{t}} > 0,$$

(A.32)

$$c = \frac{1}{(\bar{x}\bar{y} + 1)\bar{y}\bar{t}} > 0, d = \frac{\ln(1 + 1/\bar{x}\bar{y})}{\bar{t}^2} > 0.$$

(A.33)

Notation

Some specific sets

$\mathbb{R}$	Real numbers
$\mathbb{R}^n$	Real n-vectors
$\mathbb{R}^{n \times m}$	Real $m \times n$ matrices
$\mathbb{R}_+$	Non-negative real numbers
$\mathbb{R}_{++}$	Positive real numbers
$\mathbb{C}$	Complex numbers
$\mathbb{C}^n$	Complex n-vectors
$\mathbb{C}^{n \times m}$	Complex $m \times n$ matrices

Vectors and matrices

x	A scalar
$\mathbf{x}^n$	A vector with n-dimensions
$\mathbf{x}^{n \times m}$	A matrix with $n \times m$-dimensions
$\mathbf{0}$	Vector (matrix) with all components zero
$\mathbf{1}$	Vector (matrix) with all components one
$\mathbf{I}$	Identify matrix
$\mathbf{x}^T$	Transpose of matrix $\mathbf{X}$
$\mathbf{x}^H$	Hermitian (complex conjugate) transpose of matrix $\mathbf{X}$
$\lambda_i(\mathbf{X})$	ith largest eigenvalue of symmetric matrix $\mathbf{X}$
$\lambda_{max}(\mathbf{X}), \lambda_{min}(\mathbf{X})$	Maximum, minimum eigenvalue of symmetric matrix $\mathbf{X}$
$\sigma_{max}(\mathbf{X}), \sigma_{min}(\mathbf{X})$	Maximum, minimum singular value of matrix $\mathbf{X}$
$\mathbf{X}^\dagger$	Moore-Penrose or pseudo-inverse of matrix $\mathbf{X}$
$\text{diag}(\mathbf{X})$	Diagonal matrix with diagonal entries $x_1, ..., x_n$
$\text{rank}(\mathbf{X})$	Rank of matrix $\mathbf{X}$

Algebra calculators

$\lvert \cdot \rvert$	Absolute value
$\lVert \cdot \rVert$	A norm
$\mathrm{Tr}(\mathbf{X})$	Trace of matrix $\mathbf{X}$
$\lVert \mathbf{x} \rVert_1$	ℓ_1-norm of vector $\mathbf{x}$
$\lVert \mathbf{x} \rVert_2$	Euclidean (ℓ_2) norm of vector $\mathbf{x}$
$\lVert \mathbf{x} \rVert_\infty$	ℓ_∞-norm of vector $\mathbf{x}$
$\lVert \mathbf{X} \rVert_2$	Spectral norm (maximum singular value) of matrix $\mathbf{X}$
$\langle \mathbf{x}, \mathbf{y} \rangle$	Inner product of vector x and y
$\displaystyle\sum_{k=1}^{K} x$	Sum of all components in vector $\mathbf{X}$
$\displaystyle\prod_{k=1}^{K} x$	Product of all components in vector $\mathbf{X}$
$\mathbf{x} \preceq \mathbf{y}$	Componentwise inequality between vectors $\mathbf{x}$ and $\mathbf{y}$
$\mathbf{x} \prec \mathbf{y}$	Strict componentwise inequality between vectors $\mathbf{x}$ and $\mathbf{y}$
$\mathbf{X} \preceq \mathbf{Y}$	Matrix inequality between symmetric matrices $\mathbf{X}$ and $\mathbf{Y}$
$\mathbf{X} \prec \mathbf{Y}$	Strict matrix inequality between symmetric matrices $\mathbf{X}$ and $\mathbf{Y}$

Functions and derivatives

$\mathbf{dom}\, f$	Domain of function f
$f : A \to B$	f is a function on the set $\mathbf{dom}\, f \subseteq A$ into the set B
$\nabla_x f(x)$	First derivative (gradient) of function f
$\nabla^2_{xx} f(x)$	Second derivative (Hessian) of function f
$\mathbf{x} \sim \mathcal{CN}(\bar{x}, \mathbf{R}_x)$	A Gaussian random vector with mean $\bar{x}$ and covariance $\mathbf{R}_x$

Index

activation functions, 83
additive white Gaussian noise
(AWGN), 134, 159, 160
ADMM: *see* alternating direction
method of multipliers (ADMM)
air-to-air (ATA) channel, 133–4
air-to-ground (ATG) channel, 97, 127,
128–9, 128, 133–4
alternating direction method of
multipliers (ADMM), 71–3
ANN: *see* artificial neural network
(ANN)
ant colony optimisation (ACO), 41
application programming interface
(API), 75
artificial bee colony (ABC), 41
artificial intelligence (AI), 77
artificial neural network (ANN), 78
asynchronous real-time scenarios, 25

beamforming-based spatial precoding
method, 42
beamforming design, 20
best-effort systems, 26
big data, 44–5
block coordinate descent (BCD)
procedure, 137
branch-and-bound (BnB) methods, 48

cell-free networks, 22, 114–17
channel state information (CSI)
feedback, 42
Cholesky factorisation, 43

clustering UAV-GS networks (CUN),
149–51
code generation, 35
COgnition-BAsed NETworkS
(COBANETS), 79
cognitive radio networks (CRNs),
22, 79
unmanned aerial vehicles in
channel model, 158–9
problem formulation, 160–1
system model, 157–8
transmission scheme, 159–60
combinatorial optimisation, 47–8
completion time, optimal
by clustering UAV-GS networks
(CUN), 149–51, 152
by peer-to-peer UAV-GS networks,
147–9, 152
computation, 43–4
computational complexity, 40
computer processing unit (CPU), 91
constrained K-means clustering (CKC)
method
number of clusters estimation, 137
preliminaries of, 136
with QoS constraints, 136–7
constrained optimisation, 44
conventional optimisation approach,
161–2
convex functions, 1–2
convexity propositions, 11–12
Convex.jl (CVX.jl), 94
convex optimisation
classes, 12–13
geometric programming, 15–16

www.ingramcontent.com/pod-product-compliance
Lightning Source LLC
Chambersburg PA
CBHW050517190326
41458CB00005B/1564